T0135957

TECHNISCHE UNIVERSITÄT MÜNCHEN

Lehrstuhl für Biofunktionalität der Lebensmittel

Structure-related anti-inflammatory mechanisms
of probiotic bacteria

Gabriele Hörmannsperger

Vollständiger Abdruck der von der Fakultät Wissenschaftszentrum Weihenstephan für Ernährung, Landnutzung und Umwelt der Technischen Universität München zur Erlangung des akademischen Grades eines

Doktors der Naturwissenschaften

genehmigten Dissertation.

Vorsitzender: Univ.-Prof. Dr. rer. nat. M. Schemann
Prüfer der Dissertation:

1. Univ.-Prof. Dr. rer. nat. D. Haller

2. Univ.-Prof. Dr. rer. nat. S. Scherer

3. Univ.-Prof. Dr. rer. nat. A. Parlesak (schriftl. Beurteilung),
Technical University of Denmark

Die Dissertation wurde am 25.11.2009 bei der Technischen Universität München eingereicht und durch die Fakultät Wissenschaftszentrum Weihenstephan für Ernährung, Landnutzung und Umwelt am 29.1.2010 angenommen

Bibliografische Information der Deutschen Nationalbibliothek

Die Deutsche Nationalbibliothek verzeichnet diese Publikation in der
Deutschen Nationalbibliografie; detaillierte bibliografische Daten sind
im Internet über http://dnb.d-nb.de abrufbar.

ISBN 978-3-8325-2581-1

Logos Verlag Berlin GmbH
Comeniushof, Gubener Str. 47,
10243 Berlin
Tel.: +49 (0)30 42 85 10 90
Fax: +49 (0)30 42 85 10 92
INTERNET: http://www.logos-verlag.de

Für meine Familie

PUBLICATIONS AND PRESENTATIONS

Peer-reviewed original manuscripts and reviews

I.   **Hörmannsperger, G.**, Clavel, T., Hoffmann, M., Reiff, C., Kelly, D., Loh, G., Blaut, M., Hölzlwimmer, G., Laschinger, M., Haller, D. (2009).
"Post-translational Inhibition of IP-10 Secretion in IEC by Probiotic Bacteria: Impact on Chronic Inflammation"
PLoS One 4(2): e4365. doii:10.1371/journal. pone. 0004365

II. Werner, T., **Hoermannsperger, G.**, Schuemann, K., Hoelzlwimmer, G.,Tsuji, S., Haller, D. (2009).
„Intestinal epithelial cell proteome from wild-type and TNFDeltaARE/WT mice: effect of iron on the development of chronic ileitis"
J Proteome Res 8, 3252-64.

III. Reiff, C., Delday, M., Rucklidge, G., Reid, M., Duncan, G., Wohlgemuth, S., **Hörmannsperger, G.**, Loh, G., Blaut, M., Collie-Duguid, Haller, D., E., Kelly D. (2009).
"Balancing inflammatory, lipid, and xenobiotic signaling pathways by VSL#3, a biotherapeutic agent, in the treatment of inflammatory bowel disease"
Inflamm Bowel Dis, 15(11):1721-36

IV. **Hörmannsperger G.**, Haller D. (2009)
Review: Molecular crosstalk of probiotic bacteria with the intestinal immune system: Clinical relevance in the context of IBD
Int J Med Microbiol., ahead of print

V.   Sanders, M. E., Akkermans, L., Haller, D., Hammerman, C., Heimbach J., Huys, G., Levy, D., Mack, D., Phoukham, P., Constable, A., Solano-Aigula, G., Vaughan, E., **Hörmannsperger, G.**, Lutgendorff, F.
Review: Safety of probiotics
Gut Microbes, submitted

Published abstracts

1. **Hörmannsperger G.**, Clavel T., Haller D. (2008)
   Post-translational inhibition of IP-10 protein secretion through ERK-dependent and ubiquitine-mediated mechanisms: Bacterial strain specific effects of VSL#3
   Gastroenterology, Volume 134, Issue 4, Supplement 1 (A-153-A-154)

Oral presentations

1. **Hörmannsperger G.**, Clavel T., Haller D. Probiotic VSL#3 bacteria selectively inhibit TNF-induced proinflammatory cytokine expression in intestinal epithelial cells. 44. Wissenschaftlicher Kongress der Deutschen Gesellschaft für Ernährung, Halle-Wittenberg, 8.-9. March 2007

2. **Hörmannsperger G.**, Clavel T., Haller D. Probiotic VSL#3 bacteria inhibit TNF-induced IP-10 production in IEC but do not reduce TNF-induced ileitis in TNFdeltaARE-mice. 13th International Congress of Mucosal Immunology, Tokyo, Japan, 9.-12. July 2007

3. **Hörmannsperger G.**, Reiff C., Loh G., Blaut M., Kelly D., Haller D. Bacteria-host interaction under conditions of chronic inflammation: protective role of probiotic bacteria. European Nutrigenomics Organisation (NuGO) week 2007, Oslo, Norway, 18.-21. September 2007

4. **Hörmannsperger G.**, Clavel T., Haller D. Post-translational inhibition of IP-10 protein secretion in intestinal epithelial cells through ubiquitine-mediated mechanisms: bacterial strain specific effects of VSL#3. 1. Meeting der Fachgruppe "microbiota and host" der Deutschen Gesellschaft für Hygiene und Mikrobiologie, Seeon, 2.-4. Mai 2008

5. **Hörmannsperger G.**, Clavel T., Haller D. Post-translational inhibition of IP-10 protein secretion in intestinal epithelial cells through ERK-dependent and ubiquitine-mediated mechanisms: bacterial strain specific effects of VSL#3. Digestive Diseases Week 2008, San Diego, USA, 17.-22. Mai 2008

6. **Hörmannsperger G.**, Reiff C., Loh G., Blaut M., Kelly D., Haller D. Bacteria-host interaction under conditions of chronic inflammation: protective role of probiotic bacteria. European Nutrigenomics Organisation (NuGO) week 2008, Potsdam, 2.-5. September 2008

7. **Hörmannsperger G.**, Clavel T., Hoffmann M., Reiff C., Kelly D., Loh G., Blaut M., Hölzlwimmer G., Laschinger M., Haller D. Protektive Mechanismen probiotischer Mikroorganismen im Kontext chronisch entzündlicher Darmerkrankungen. 46. Wissenschaftlicher Kongress der Deutschen Gesellschaft für Ernährung, Gießen, 12.-13. March 2009

8. **Hörmannsperger G.**, Clavel T., Hoffmann M., Reiff C., Kelly D., Loh G., Blaut M., Hölzlwimmer G., Laschinger M. Haller D. Post-translational Inhibition of IP-10 Secretion in IEC by Probiotic Bacteria: Impact on Chronic Inflammation. 14th International Congress of Mucosal Immunology, Boston, USA, 5.-8. July 2009

9. **Hörmannsperger G.**, Post-translational Inhibition of IP-10 Secretion in IEC by Probiotic Bacteria: Impact on Chronic Inflammation. 6th prebiotic and probiotic conference, Rome, Italy, 13.-15. September 2009, **Invited Lecture**

Poster presentations

1. **Hörmannsperger G.**, Haller D. Probiotic bacteria inhibit TNF-induced IP-10 gene expression through the modulation of transcription factor recruitment to the chemokine promoter in intestinal epithelial cells. European Nutrigenomics Organisation (NuGO) week 2006, Oxford, UK 12.-15. September 2006

2. **Hörmannsperger G.**, Haller D. Probiotic bacteria inhibit TNF-induced IP10 expression through the modulation of transcription factor recruitment to the chemokine promoter in intestinal epithelial cells. 5th meeting of the European Mucosal Immunology Group, Prague, Czech Republik 5.-7. October 2006

3. **Hörmannsperger G.**, von Schillde M., Clavel T., Hoffmann M., Reiff C., Kelly D., Loh G., Blaut M., Hölzlwimmer G., Laschinger M., Haller D. VSL#3-derived *L. casei* induces post-translational degradation of IP-10 protein in intestinal epithelial cells: impact on chronic inflammation. 2. Meeting der Fachgruppe "microbiota and host" der Deutschen Gesellschaft für Hygiene und Mikrobiologie, Seeon 2.-4. Mai 2009

4. **Hörmannsperger G.**, Clavel T., Hoffmann M., Reiff C., Kelly D., Loh G., Blaut M., Hölzlwimmer G., Laschinger M., Haller D. VSL#3-derived *L. casei* induces post-translational degradation of IP-10 protein in intestinal epithelial cells: mpact on chronic inflammation. European Nutrigenomics Organisation (NuGO) week 2009, Montecatini, Italy 31. August – 3.September 2009

Grants

1. Travel grant from the European Mucosal Immunology Group for the 5th meeting of the European Mucosal Immunology Group, Prague, Czech Republic, 5.-7. October 2006

2. Travel grant from Deutsche Forschungsgesellschaft (DFG) for the 13th International Congress of Mucosal Immunology, Tokyo, Japan, 9-12. July 2007

3. Travel grant from GlaxoSmithKline for the Digestive Diseases Week 2008, San Diego, USA, 17.-22. Mai 2008

4. Travel grant from Boehringer Ingelheim to finish a collaborative experimental study in the group of Prof. Balfour Sartor at the University of North Carolina, Chapel Hill, USA and a subsequent lectureship in the group of Prof. Brent Polk at the Vanderbilt University, Nashville, USA, 7.-19. September 2008

5. Travel grant from Danone Institute for the 14th International Congress of Mucosal Immunology, Boston, USA, 5.-8. July 2009

Rewards

Yakult young scientist award "Science for Health",
5th International Yakult Symposium, „The Gut and More – Probiotic influences beyond the gut", Amsterdam, 18.-19. June 2009

ZUSAMMENFASSUNG

Chronisch entzündliche Darmerkrankungen (CED) sind immunvermittelte Erkrankungen des Gastrointestinaltraktes, die durch eine Störung des Darmgleichgewichtes charakterisiert sind. Obwohl die intestinale Mikrobiota bei CED eine wichtige proinflammatorische Rolle einnimmt, führt die orale Aufnahme von VSL#3, einer probiotischen Mischung aus acht verschiedenen Milchsäurebakterienstämmen in klinischen Studien zu einer Reduktion der Entzündung. Im Kontrast zu der klinischen Relevanz sind die antientzündlichen Wirkstrukturen und Wirkmechanismen von VSL#3, welche der Stärkung des Darmgleichgewichtes zugrunde liegen, weitgehend unbekannt. Das intestinale Epithel ist als Schnittstelle zwischen den luminalen Antigenen und dem intestinalen Immunsystem maßgeblich an der Aufrechterhaltung des Darmgleichgewichts beteiligt. Die intestinalen Epithelzellen (IEC) verarbeiten die Signale von beiden Seiten und tragen zur Induktion einer adequaten Immunantwort bei. Diese immunregulierende Funktion der IEC ist bei CED gestört. Das Ziel der vorliegenden Arbeit war somit die Charakterisierung protektiver Wirkmechanismen von VSL#3 auf der Ebene der IEC sowie die Identifizierung protektiver bakterieller Strukturen.

In vitro führt die Stimulation von aktivierten IEC mit VSL#3 zu der selektiven Reduktion eines entzündungsfördernden Chemokins, des Interferon-induzierbaren Proteins (IP)-10. Dieser protektive Effekt wird durch Lactocepin, eine zellwandständige und sekretierte Serinprotease von *Lactobacillus casei*, vermittelt. Lactocepin induziert eine IP-10-spezifische Sekretionsblockade und in der Folge den Abbau von IP-10 in IEC. Auf funktioneller Ebene blockiert die selektive Reduktion von IP-10 die IEC-induzierte T-Zell-Transmigration. Fütterungsstudien in Mausmodellen für chronische Ileitis (TNF$^{\Delta ARE/+}$) und chronische Colitis (IL-10-/-) zeigten, dass die antientzündliche Wirksamkeit von VSL#3 mit der Reduktion von IP-10 in den IEC korreliert und je nach Lokalisation der Entzündung im Darm variiert.

Die Ergebnisse der vorliegenden Arbeit deuten darauf hin, dass die Lactocepin–induzierte Inhibierung der IP-10 Sekretion in IEC einen klinisch relevanten probiotischen Mechanismus im Kontext spezifischer IBD Indikationen darstellt. Die Identifizierung von Lactocepin als eine der ersten probiotischen Wirkstrukturen könnte einen wichtigen Beitrag zur Entwicklung neuartiger Therapiemethoden leisten.

ABSTRACT

Inflammatory bowel diseases (IBD) are immune-mediated chronic diseases which are characterized by an overreaction of the intestinal immune system towards the intestinal microbiota. VSL#3, a mixture of eight different lactic acid bacteria, is a clinically relevant probiotic compound in the context of IBD but the bacterial structures and molecular mechanisms underlying the observed protective effects on the intestinal homeostasis are largely unknown. The intestinal epithelium plays a very important role in the maintenance of the intestinal homeostasis since the intestinal epithelial cells (IEC) are capable of sensing signals from the luminal microbiota and the intestinal immune system. IEC process the signals from both sides and elicit an adequate immune response. In IBD, this important immune regulatory functions of the IEC is lost due to dysregulated activation of the IEC, resulting in increased secretion of immune-activating chemokines and cytokines. Thus, the aim of the present study was to reveal protective mechanisms of VSL#3 on IEC function and to identify active probiotic structures.

In vitro, VSL#3 was found to selectively inhibit activation-induced secretion of the T-cell chemokine interferon-inducible protein (IP)-10 in IEC. Lactocepin, a cell-wall associated and secreted serine protease of VSL#3-derived *Lactobacillus casei* was identified to be the active anti-inflammatory component of VSL#3. Mechanistically, *Lactobacillus casei* did not impair initial IP-10 protein production but induced an IP-10-specific secretory blockade followed by subsequent degradation of the proinflammatory chemokine in IEC. Specific inhibition of IP-10 secretion was found to be functionally relevant as it resulted in abrogation of IEC-mediated T-cell transmigration. VSL#3 feeding studies in mouse models for experimental ileitis (TNF$^{\Delta ARE/+}$) and experimental colitis (IL-10-/-) revealed that VSL#3 has intestinal compartment specific protective effects on the development of inflammation. Reduced histopathological inflammation in the cecum of IL-10-/- mice after VSL#3 treatment was found to correlate with reduced levels of IP-10 protein in primary cecal epithelial cells.

The results of the present thesis suggest that the inhibitory effect of lactocepin on IP-10 secretion in IEC is an important probiotic mechanism that contributes to the anti-inflammatory effects of VSL#3 in specific subsets of IBD patients. Most important, the identification of lactocepin as one of the first protective probiotic structures can serve as the basis for the development of new efficient therapeutical strategies in the context of IBD.

TABLE OF CONTENTS

1 INTRODUCTION

Bacteria: from foe to friend.

Bacteria were discovered and described for the first time by Antoni van Loewenhoek in the 17th century. From then on, bacteria were found to be the cause of many, often life-threatening infectious diseases and generally regarded as a threat towards human health. The finding that the human intestinal tract harbours large amounts of these microorganisms in the 19th century was thereby thought to be a disease-associated situation and was called "intestinal toxaemia". In logical consequence, the british royal surgeon Sir William Arbuthnot Lane (1856-1943) suggested to remove the large intestine in order to get rid of these dangerous inhabitants.

Some decades later, it was found that the normal intestinal microbiota is not only harmless but even necessary for the development and protection of human intestinal function and health. In addition, oral uptake of specific bacterial strains was found to exert beneficial effects in the context of intestinal disorders like diarrhea. This finding constitutes the origin of the probiotic theory and led to extensive investigations of numerous bacterial strains in order to reveal their potential to prevent or reduce a variety of intestinal and extra-intestinal diseases. Although several clinical and experimental studies proved beneficial effects of different bacterial strains in various disease indications, these scientific studies revealed that probiotic efficacy and safety is highly dependent on a complex interplay between the bacterial strain, the resident intestinal microbiota, specific disease indications and general health of the host as well as on timing and dosage of the probiotic treatment. These findings led to the current World Health and Food and Agriculture Organisation (WHO/FAO) definition of probiotics as "living organisms, which, after oral uptake in sufficient amounts, exert a beneficial effect on the host".

With regard to this definition, almost none of the "probiotic" food supplements and functional food products, that are currently sold at the supermarket, do contain probiotic bacteria. In spite of the fact that the claimed health benefits have not been proven yet, the giantic, constantly increasing market volume of these "probiotic" products clearly shows that the view of bacteria turned around from foe to friend.

1.1 Microbiota-host interactions in health and disease

1.1.1 Development and complexity of the microbial ecosystem in the gastrointestinal tract.

The intestinal tract of an adult individual harbours 10^{14} cfu bacteria in total, a number that exceeds the amount of our "own" cells by a factor of 10. Every individual is born sterile but gets rapidly colonized from the moment of birth and develops a characteristical "fingerprint" microbiota which is stable during adulthood until it gets more variable in elderly people again (Woodmansey, 2007). The composition of this "adult" intestinal microbiota was found to be highly dependent on the very first microbes colonizing the intestine after birth. Initial colonisation of the gastrointestinal tract is characterized by the presence of facultative anaerobic or aerotolerant bacteria which acidify and deplete oxygen in the intestine, thereby setting the stage for the colonization with anaerobic bacteria which are then the dominating group of the adult microbiota (Mackie et al., 1999). In this context, it has been shown that the microbiota of caesarean born children differs from the one of vaginal birth children (Harmsen et al., 2000). The fact that the former ones have an increased risk of developing allergic diseases later in life (Wu and Chen, 2009) might be a consequence of these delivery-dependent changes in the intestinal microbiota.

Regarding the variability of bacteria in other ecological niches, it is quite surprising that the intestinal microbiota of an adult individual mainly consists of only four phyla: Firmicutes, Bacteroidetes, Actinobacteria and Proteobacteria. These phyla show extremely high diversity on the subgenus level (Backhed et al., 2005). Culture based methods to determine the intestinal diversity resulted in estimations of about 400-500 different species (Moore and Holdeman, 1974). The development of molecular techniques to identify bacteria, like the analysis of 16S ribosomal RNA, led to assessments reaching from approximately 1000 (Eckburg et al., 2005) to about 15000 (Frank et al., 2007) different species in the intestine. This discrepancy between the results of culture-based and molecular methods clearly shows that the vast majority of the intestinal bacterial species are uncultivable, making it difficult to describe these bacteria or to investigate their functions.

The gastrointestinal tract of a healthy individual offers various different niches for bacteria, resulting in a characteristical composition of the intestinal microbiota in different intestinal segments. Bacterial numbers continuously increase from the almost sterile stomach (10^3 cfu/g) to the highest load of mostly anaerobic bacteria in the large intestine (10^{12} cfu/g) (Holdeman et al., 1976) (Figure 1).

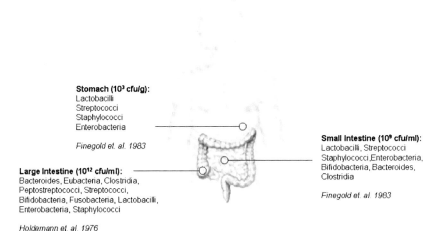

Stomach (10^3 cfu/g):
Lactobacilli
Streptococci
Staphylococci
Enterobacteria

Finegold et. al. 1983

Large Intestine (10^{12} cfu/ml):
Bacteroides, Eubacteria, Clostridia,
Peptostreptococci, Streptococci,
Bifidobacteria, Fusobacteria, Lactobacilli,
Enterobacteria, Staphylococci

Holdemann et. al. 1976

Small Intestine (10^9 cfu/ml):
Lactobacilli, Streptococci
Staphylococci,Enterobacteria,
Bifidobacteria, Bacteroides,
Clostridia

Finegold et. al. 1983

Figure 1. The diversity and number of intestinal microbes increases from the stomach to the distal part of the gastrointestinal tract.

The intestinal microbiota is characterized by high interindividual variability, although the application of molecular methods recently proved the existence of a subset of bacterial species which are present in most healthy individuals, the so-called core microbiome (Tap et al., 2009). In addition, the intestinal microbiota of a single individual is prone to fluctuations dependent on the lifestyle, environment and health status of the host. In this context, it is important to understand that it is not only the composition of the intestinal microbiota, but the functional activity of every single constituent at a given point of time, that has an impact on the intestinal health and *vice versa*.

The intestinal microbiota fulfills various important functions like providing the host with additional energy by fermentation of undigestible food components or the production of important micronutrients like vitamin B or K. Furthermore, the intestinal microbiota is involved in bile acid metabolism and influences the activity of xenobiotics and secondary plant metabolites (Cummings and Macfarlane, 1997). Changes in the composition and functionality of the intestinal microbiota were associated with many intestinal disorders (Tannock, 2008) but also extraintestinal diseases like allergy (Sepp et al., 2005), liver diseases (Wigg et al., 2001) and adipositas (Kalliomaki et al., 2008). These findings indicate that the intestinal microbiota might be an important target for therapeutic interventions in the context of these diseases. However, a lot of research needs to be done in order to get a grip on the complex environmental and host factors influencing the composition and function of the intestinal microbiota. This knowledge would possibly enable to modulate the intestinal microbiota in a targeted way in the future.

1.1.2 Impact of the intestinal microbiota on the intestinal immune system.

The importance of the intestinal microbiota on gut health and subsequently on the overall health of an individual was not experienced until antibiotics were broadly used in the treatment of infectious diseases from the 1950s on. Antibiotics were unexpectedly found to have detrimental effects on the host via their negative impact on the intestinal microbiota. Antibiotic-associated diarrhea and infections gave a first hint that the intestinal microbiota is pivotal for the gastrointestinal health of the host. Today, it is common knowledge that the resident intestinal bacteria prevent the colonisation and attachment of pathogens via the occupation of mucosal attachement sites, the secretion of bacteriocins, the reduction of the intestinal pH and the competition for micronutrients (Levy, 2000).

Apart from these intrinsic pathogen restrictive functions of the intestinal microbiota, studies with gnotobiotic (germfree) animals showed that the signals derived from intestinal bacteria are necessary for the development of a morphologically and functionally normal intestinal immune system. Under normal circumstances, the intestinal immune system is able to sense intestinal bacteria and their products via pattern recognition receptors (PRR), mainly Toll like receptors (TLR) or cytosolic nucleotide binding oligomerization domain (NOD) receptors, and reacts upon these signals in an adequate way (Medzhitov, 2001; Medzhitov and Janeway, 2000). The absence of the intestinal microbiota and microbiota-derived signals in germfree animals was found to result in abnormal immune development. Germfree animals show a strongly enlarged cecum and reduced vascularisation of the lamina propria as well as disturbed differentiation of IEC. Furthermore, germfree mice were found to be immunodeficient due to the lack of functional Peyer's Patches as well as reduced presence of intraepithelial and lamina propria lymphocytes (Thompson and Trexler, 1971). In addition, the development of B and T cell follicles in mesenteric lymph nodes and even in the spleen was found to be impaired in germfree animals (Gordon, 1959). Compared to conventional mice, germfree mice were found to be biased towards T helper (h) 2-type immune activity resulting in impaired induction of oral tolerance mechanisms (Sudo et al., 1997). Early colonisation of germfree mice with normal intestinal microbiota, single bacterial strains or even specific single bacterial antigen like the polysaccharide A (PSA) of *Bacteroides fragilis* were found to result in normalization of the aberrant intestinal morphology and immune function (Hooper et al., 2001; Mazmanian et al., 2005; Sudo et al., 1997).

These results clearly show the importance of microbiota-derived signals especially during immune maturation for the development of an efficient host defense system which is able to discriminate between harmless and pathogenic antigens. The intestinal immune system must tolerate huge amounts of highly variable food- and commensal-derived antigens in order to prevent constant inflammation whereas it needs to prevent and cease infections. It is still largely unknown how the intestinal immune system is able to differentiate between harmless microbes and pathogens, as both bacterial groups express similar microbial associated molecular pattern (MAMP), which are called pathogen associated molecular pattern (PAMP) in the context of pathogens. However, pathogenic bacteria induce strong inflammatory immune responses whereas harmless microorganisms are tolerated by the intestinal immune system. In this context, immune tolerance towards ingested *Lactobacillus (L.) plantarum* was found to be a consequence of transient activation of proinflammatory pathways in the duodenal mucosa of healthy humans (van Baarlen et al., 2009), suggesting this kind of physiological inflammation to play an important role in the induction of tolerance towards newly encountered antigens. These homeostatic mechanisms are abrogated in IBD patients, resulting in chronic inflammation of the gastrointestinal tract.

1.1.3 Disturbed microbiota-host interactions in IBD.

About 0.1% of the population of Europe, North America and Japan suffer from one of the two idiopathic IBD disorders, Crohn´s Disease (CD) or Ulcerative colitis (UC), whereas these diseases are almost unknown in third world countries (Loftus, 2004). The fact that the incidence rate of IBD increases in parallel to progressing industrialization indicates that the environment plays a very important role in the development of IBD. Although the initial trigger inducing the development of IBD is unknown, the intestinal microbiota was revealed to be one of the environmental factors that constitute a major proinflammatory drive in genetically susceptible hosts (Sartor and Muehlbauer, 2007).

The microbiota of IBD patients was found to differ in composition and metabolic activity from the one of healthy individuals, and the amount of mucosa-associated bacteria was found to be increased in IBD (Sartor, 2008). In addition, the reduction of the intestinal microbiota by antibiotic treatment (Schultz et al., 2003) or by diversification of the fecal stream (Ginzburg et al., 1939; Winslet et al., 1994) results in reduced disease activity. Furthermore, a meta-analysis of genome-wide association studies defined more than 30 susceptibility loci for CD (Barrett et al., 2008), mostly affecting genes that are responsible for bacterial recognition and clearance (NOD2/CARD15, ATG16L1, IRGM) as well as immune regulation (IL23R, IL12B) (Cho, 2008). This finding supports the impact of dysregulated bacteria-host interactions in IBD.

Some of these gene polymorphisms, like IL23R, or additional ones, like MAGI2, a protein important for the mucosal barrier maintenance (McGovern et al., 2009), were also found to play a role in UC. Generally, genetic susceptibility seems to be more important in CD as the disease concordance rate between monozygotic twins is significantly higher in CD than in UC (Tysk et al., 1988). The two idiopathic IBD disorders differ with regard to disease location and pathology. CD is a Th1-mediated transmural inflammation which can affect every part of the gastrointestinal tract and lead to extraintestinal manifestations like fistulas. In contrast, UC is characterized by a mixed Th1/Th2-type inflammation and is restricted to the mucosa of the colon.

However, CD as well as UC are characterized by loss of tolerance of the intestinal immune system towards the intestinal microbiota (Figure 2). The loss of tolerance leads to constant immune activation followed by intestinal inflammation, mucosal tissue damage and reduced intestinal barrier function. In consequence, penetration of more immune-activating antigens and pathogens into the host tissue is facilitated inducing further progression of the inflammatory response. This viscious circle is thought to be due to a complex interplay between changes of the intestinal microbiota, disturbances of IEC function and dysregulated immune cell responses, resulting in the chronic disease state that is characteristic for IBD.

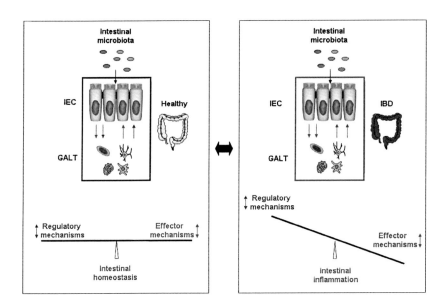

Figure 2. Loss of intestinal homeostasis results in chronic intestinal inflammation.
Microbiota-derived signals induce tolerogenic mechanisms in a healthy host, resulting in intestinal homeostasis. In contrast, constant IEC activation and overshooting proinflammatory effector mechanisms in response to the intestinal microbiota result in the development of IBD. Abbreviations: GALT; gut associated lymphoid tissue

1.2 IBD are characterized by disturbed IEC functions and dysregulated immune cell activities.

1.2.1 Impaired intestinal epithelial barrier functions in IBD.

The intestinal epithelium is a cell monolayer that constitutes a selective barrier between luminal contents and the underlying lamina propria immune cells. The intestinal epithelial layer is the site of efficient and selective uptake, transport and intercellular penetration of digested nutrients but luminal microorganisms and bacterial antigens are effectively excluded from host tissues. Reduced or abrogated intestinal epithelial barrier function was found to be prominent in IBD patients but it is not known whether this is a primary defect preceding IBD development or whether reduced intestinal barrier function is a consequence of ongoing inflammation and tissue disruption.

Physical epithelial barrier function is dependent on the integrity of the epithelial cell monolayer which is determined by IEC viability and the spatial expression of tight junctions (TJ) between neighbouring IEC. In this context, continuous supply of fresh and undamaged IEC by pluripotent stem cells at the bottom of the crypts is pivotal for proper IEC barrier function. The controlled high IEC turnover constitutes an effective barrier mechanism as even if pathogens succeed in attaching or infecting IEC, these cells are often shedded off and replaced before the pathogens can proliferate and spread into deeper host tissues. In contrast, increased IEC apoptosis, as it was observed in inflamed tissue of IBD patients, is highly detrimental on epithelial barrier function (Strater et al., 1997). Apart from infection- or inflammation-mediated IEC apoptosis, reduced IEC viability could be due to disturbed IEC energy homeostasis in IBD (Roediger, 1980). Constitutive high energy consumption as a result of persistent IEC activation on the one hand and reduced energy supply due to impaired nutrient digestion, uptake and bioavailability on the other hand most probably contribute to the failure of IEC to maintain energy homeostasis in IBD. In the context of reduced nutrient bioavailability, the level of short chain fatty acids (SCFA), which are mainly produced by Bacteroides and Clostridia species in the healthy colon, was found to be decreased in IBD patients (Marchesi et al., 2007). SCFA constitute a major source of energy for enterocytes (Gassull, 2006) and impaired energy availability was shown to be a major inductor of endoplasmic reticulum (ER) stress responses (Zhao and Ackerman, 2006). In contrast to transient ER stress, which is a homeostatic response resolving cellular stress situations like an accumulation of misfolded proteins, prolonged ER stress has deleterious effects and induces IEC apoptosis (Messlik et al., 2009; Shkoda et al., 2007).

The intestinal epithelial barrier in IBD patients was also found to be impaired due to reduced expression and redistribution of TJ proteins (Schulzke et al., 2009), indicating facilitated paracellular entry of proinflammatory antigens and infectious agents into the host tissue.

Apart from constituting a mere physical barrier, the intestinal epithelium is armed with various biochemical and immunological means in order to prevent bacterial attachment and translocation. The pluripotent stem cells at the bottom of the crypts give rise to four differently specialized IEC types which contribute to the intestinal barrier via unique features. The majority of IEC are enterocytes, which, besides their nutrient absorptive function, are continuously transporting plasma cell derived dimeric immune globulin (Ig) A into the intestinal lumen. The binding of IgA to its respective antigen in the lumen is an important mean to neutralize toxins, to prevent pathogenic adherence and translocation as well as antigen penetration into the mucosa. In IBD patients, the concentration of IgA in fecal content was found to be strongly reduced indicating a defect in this important first-line defense system (Brandtzaeg et al., 2006).

In addition to the transport of IgA, enterocytes do produce and secrete anti-microbial β-defensins. Defensins are small cationic proteins which are able to bind to bacterial surfaces and to kill bacteria by forming pores into the bacterial cell wall. The main producers of defensins are Paneth cells, which are mainly located at the bottom of the crypts and produce high levels of antimicrobial α-defensins (Salzman et al., 2007) in order to protect the stem cells. In IBD patients, defensin production from enterocytes and paneth cells has been shown to be strongly reduced (Ramasundara et al., 2009). Furthermore, the variation of the CD susceptibility gene NOD2/CARD15 is strongly associated with reduced secretion of human defensin 5 (HD-5) by ileal Paneth cells. These data generated the hypothesis that CD ileitis may be in part a defensin deficiency syndrome (Wehkamp et al., 2005a; Wehkamp et al., 2005b) but although this finding shows that microbial clearance is hampered in IBD, it is still unclear whether reduced defensin production is a cause or consequence of IBD.

A highly viscous mucus layer on top of the IEC monolayer constitutes the very first physical barrier for luminal antigens. The mucus is produced by Goblet cells and was shown to strongly reduce direct contact between IEC and intestinal microbes. The inflamed tissue of UC patients is characterized by loss of Goblet cells and consistently, biopsies of UC and CD patients revealed impaired mucin production compared to controls (Moehle et al., 2006). Furthermore, missense mutations of a gene coding for a specific mucin (Muc2) were found to result in spontaneous development of colitis in transgenic mice (Heazlewood et al., 2008), suggesting that the observed defective mucin production in humans might be a primary barrier defect contributing to the development of IBD.

The fourth epithelial cell type, the hormone-secreting enterochromaffine cells, are known to be in close proximity to enteric nerves but whether they contribute to the intestinal epithelial barrier function remains to be elucidated.

Although it is commonly accepted that impaired intestinal barrier function does promote intestinal inflammation, it was recently found that the reduction of the intestinal barrier via constitutively active myosine light chain kinase is not sufficient to induce spontaneous colitis in mice (Su et al., 2009). This result clearly shows that additional disease promoting factors are necessary for the development of chronic intestinal inflammation. In this context, it is important to know that disturbances of IEC function were found to play an important role in the loss of intestinal immune homeostasis in IBD

1.2.2 Loss of IEC homeostasis in IBD.

IEC constitute an important component of the intestinal immune system as they were found to be capable to sense and respond to the intestinal microbiota as well as to directly interact with lamina propria dendritic cells (Niess et al., 2005), lamina propria lymphocytes and intraepithelial lymphocytes (Neutra et al., 2001). Dependent on the kind, amount and localisation of the bacterial stimuli, PRR activate intracellular signalling pathways in IEC which result in subsequent induction of adequate immune reactions (Artis, 2008).

In a healthy host, IEC are in a relatively unsensitive and unreactive state towards the normal luminal microbiota. It is thought that TLRs are only expressed at low numbers and in a compartimentalized way in IEC (Abreu et al., 2003; Otte et al., 2004). In addition, microbiota-derived signals that initiate the activation of proinflammatory NFκB or mitogen activated protein kinases (MAPK)-dependent pathways via PRR are ceased by intrinsic negative regulatory mechanisms of these pathways in IEC (Haller and Jobin, 2004) or by host-derived regulatory signals (Haller, 2006). Transient IEC activation after the encounter of new harmless antigens is even thought to play a role in the induction of homeostatic mechanisms as monocolonisation of germfree rats with non-pathogenic *Bacteroides vulgatus* did induce transient RelA phosphorylation (Haller et al., 2002). Even the epithelium of mice with an established microbiota was found to be sensitive to new luminal microbes. The reconstitution of lactobacilli-free mice with probiotic *L. reuteri* resulted in transient IEC activation but without any pathological consequences (Hoffmann et al., 2008). The fact that the CD susceptibility polymorphism NOD2 3020insC is associated with defective bacterial recognition and failure to activate NFκB in IEC (Barnich et al., 2005), suggests NOD2-dependent NFκB activation to play an important role in the maintenance of intestinal homeostasis.

In contrast to the induction of tolerance towards harmless antigens, pathogenic attachment or infection results in strong and enduring activation of NFκB and MAPK pathways in IEC. In addition to pathogenic stimuli, lymphocyte or leukocyte-derived proinflammatory mediators like TNF and IFNγ strongly activate NFκB as well as Janus Kinase (JAK)/signal transducer and activator of transcription (STAT) dependent pathways in IEC (Figure 3). The major consequence of the activation of these pathways in IEC is the upregulation of PRR (Abreu et al., 2002) as well as the production and secretion of cytokines and chemokines.

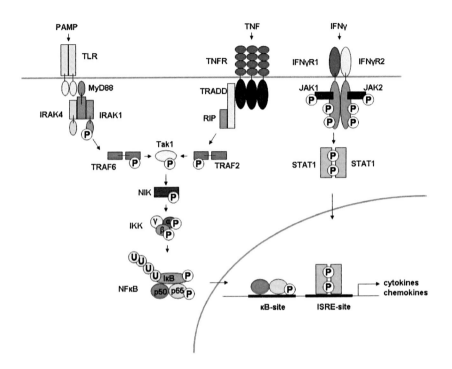

Figure 3. PAMPs and the immune cell-derived cytokines TNF and IFNγ are potent inducers of NFκB and STAT1-dependent cytokine and chemokine expression.
The activation of TLRs by PAMPs results in the formation of TLR homodimers and subsequent recruitment of the adaptor protein MyD88. Upon binding, MyD88 recruits IL-1 receptor associated kinase (IRAK) to the receptor complex which is subsequently autophosphorylated. After its dissociation from the receptor complex it recruits the kinase TRAF-6. TRAF-6 phosphorylates transforming growth factor activated kinase (Tak) 1, a downstream kinase which is also activated by TNF-induced signalling. Tak1 in turn phosphorylates NIK, the NFκB inducing kinase which then activates the IκB kinase complex (IKK). This complex is responsible for the phosphorylation of the NFκB inhibitor IκB which is followed by subsequent ubiquitination and degradation of IκB. The freed and activated p50/p65 complex then translocates into the nucleus, binds to κB-binding sites and activates the transcription of proinflammatory cytokines and chemokines. TNF-binding induces trimerization of the TNF receptor resulting in association of the TNF receptor associated death domains (TRADD) and the receptor interacting protein (RIP) which in turn recruits and activates TRAF-2. TRAF-2 subsequently phosphorylates Tak1 resulting in the above mentioned downstream signals. IFNγ-binding induces heterodimerization of IFNγ receptors 1 and 2, which induces the activation of receptor bound JAKs. Activated JAKS in turn phosphorylate STAT1 which homodimerizes and translocates into the nucleus where it activates proinflammatory chemokine transcription at interferon signal response element binding sites (ISRE).

Upon activation with PAMPs or proinflammatory cytokines, IEC were found to secrete an array of cytokines like IL6, IL18 or IL1β, which are important for the activation of different effector cells. In addition, the secretion of various chemokines, aiming to attract innate and adaptive immune effector cells into the mucosal tissue, plays a central role in the induction of mucosal immune responses by IEC. Chemokines bind to their respective seven-transmembrane G-protein coupled receptors on the surface of target cells and induce the migration of these cells towards the site of chemokine secretion along a chemokine concentration gradient. Most chemokine receptors can bind different chemokines but every chemokine induces the transmigration of a specific subset of effector cells and characteristic proinflammatory effects (Rollins, 1997). The recruitment of monocytes by monocyte chemoattractant protein (MCP) 1 or eotaxin as well as neutrophils by IL-8 results in pathogen clearance through the production of antibacterial substances like superoxide anions, nitric oxide or proteolytic enzymes. In addition, natural killer cells as well as antigen specific complement determining (CD) 8+ T cells are recruited to kill and erade infected host cells. The attraction of T and B cells via specific chemokines is responsible for the induction of highly efficient antigen-specific mechanisms to erade extracellular or phagocytosed pathogens. In a healthy host, these proinflammatory activities are readily terminated after successful elimination of the infectious agent, thereby limiting the tissue disruptive effects of these effector mechanisms.

In contrast to the tightly regulated immune response towards harmless microbes or pathogens in a healthy host, IBD are characterized by failure of the intestinal immune system to mount an adequate immune response towards antigens (Sartor and Muehlbauer, 2007). In this context, IEC of IL10-/- mice, an experimental model of chronic colitis, were found to be persistently activated upon monocolonisation with colitogenic *Enterococcus faecalis*, whereas wildtype mice show only transient IEC activation (Ruiz et al., 2005). Unceased IEC activation results in enhanced cytokine and chemokine secretion followed by constant recruitment and activity of effector cells. Prolonged effector cell activity subsequently results in mucosal tissue damage and reduction of the intestinal barrier. Consistently, increased chemokine secretion has been shown to play an important role in the development and perpetuation of experimental inflammation (Scheerens et al., 2001) as well as in IBD patients (Singh et al., 2007) (Figure 4). In consequence, these IBD-associated chemokines, especially CX3CL1/fractalkine, CCL20/MIP3α, CCL25/TECK and CXCL10/IP-10, as well as the respective receptors, are thought to be promising new targets for the therapy of IBD (Nishimura et al., 2009).

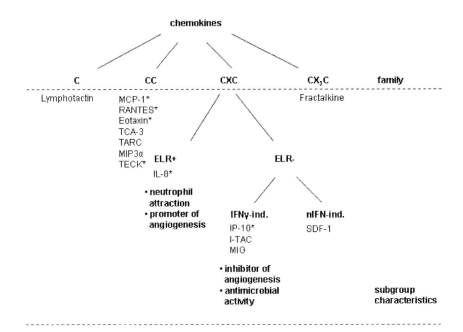

Figure 4. Chemokine classification.
Chemokines are classified into C, CC, CXC and CX3C families according to their conserved N-terminal cystein motifs with X indicating a non-conserved amino acid residue. CXC-chemokines are further subdivided according to the presence or absence of an amino-terminal amino acid sequence, the glutamine acid-leucin-arginine (ELR) motif, just before the cystein residues, which is necessary for the recruitment of neutrophils and the ability to induce or inhibit angiogenesis. In addition, the IFNγ-inducible CXC ELR- chemokines were found to have antibacterial activity (Cole et al., 2001). All chemokines that are displayed in the scheme were found to play a role in IBD and the chemokines that are marked with * were found to be secreted by IEC. Abbreviations: I-TAC; IFN-inducible T cell α attractant, MIG; monokine induced by IFNγ, MIP3α; macrophage inflammatory protein 3α, SDF-1; stromal derived factor 1, TCA-3; thymocyte-derived chemotactic agent-3, TARC; thymus and activation regulated chemokine, TECK; thymus-expressed chemokine

Apart from their role in initiating mucosal immune responses via the secretion of cytokines and chemokines, IEC were also found to be able to express major histocompatibility complex (MHC) II and non-classical Ib MHC molecules as well as costimulatory molecules upon activation, making them capable to present antigens to T cells. This finding indicates that IEC might also play a role in the induction of antigen-dependent adaptive immunity (Ruemmole et al., 1998). In summary, disturbed IEC activation and function was found to contribute to the dysregulated intestinal immune response in IBD.

1.2.3 Dysregulated activities of the intestinal immune system in response to the normal intestinal microbiota in IBD.

Uncontrolled and aggressive immune cell reactions towards antigens of the intestinal microbiota, as it is observed in IBD, can be the consequence of enhanced effector T cell activity or the failure of adequate negative regulatory mechanisms. In a healthy host, luminal food-or microbiota-derived antigens induce oral tolerance, which is characterized by the suppression of mucosal as well as systemic immune reactions towards these antigens. Oral tolerance is dependent on the induction of regulatory T cells, the induction of clonal anergy or clonal deletion of antigen-specific T cells and these mechanisms are favored by the gut associated lymphoid tissue (GALT) (Weiner, 1994). In general, the development of a naïve T cell into an effector or a regulatory T cell is dependent on the cytokine milieu and on the costimulatory activity of the antigen presenting cell (APC). In addition, the nature and amount of the presented antigen plays an important role. Secretion of IL-12, IFNγ or IL-17 by APC induces proinflammatory Th1 or Th17 effector cells, whereas antigen presentation in the presence of IL-4 induces the generation of Th2 effector cells. In contrast, the cytokines IL-10 and transforming growth factor (TGF) β mediate the development of regulatory T cells which are important for the induction of tolerance mechanisms (Doganci et al., 2005) (Figure 5). In consequence, dysregulated immune responses on the T cell level can be due to increased induction of proinflammatory immune cells by dysregulated APCs or due to intrinsically increased proliferation and activity of the respective effector cells. Most important, the induction of effector cells results in the suppression of regulatory cells and vice versa (Strober et al., 2002).

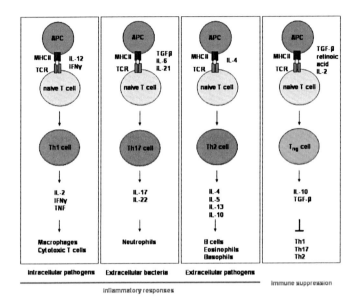

Figure 5. Development and effector functions of different T cell subsets.
The cytokine milieu in which antigen presentation to a naïve T cell takes place drives the generation of different T cell subsets. Every effector T cell subset subsequently activates specific effector cell types which contribute to the clearance of pathogens. In contrast, regulatory T cells (Treg) suppress the activity of effector T cells and induce tolerance mechanisms or reduce inflammation. Abbreviations: TCR; T cell receptor, MHCII; major histocompatibility complex II

CD was found to be a typical Th1-mediated disease showing high levels of IFNγ and TNF in the intestinal mucosa, whereas it is not that clear in UC, showing some indications for a Th2-type disease (Hanauer, 2006). However, the induction of oral tolerance towards newly encountered luminal antigens was found to be defective in both IBD indications (Kraus et al., 2004), suggesting an intrinsic failure of regulatory mechanisms in IBD. In addition, increased levels of Th17-derived IL-17 were detected in the mucosa of UC and CD patients (Fujino et al., 2003).

Interestingly, most experimental IBD models show a Th1-type disease, mimicking CD with regard to the effector mechanisms whereas the disease location is almost exclusively the colon like in UC (Table 1). The reason for this discrepancy is unclear but might be due to the higher bacterial load in the colon compared to the ileum, making it more vulnerable to disease induction. Experimental IBD models are subclassified into type I models, which develop chronic inflammation due to increased effector cell function, and type II models, which develop chronic inflammation due to disturbed regulatory mechanisms.

Table 1: Common murine IBD models grouped by disease phenotype.

Th1 Models		Th2 Models	
Colitis	**Ileitis**	**Colitis**	**Ileitis**
SCID-transfer**	Samp1/Yit	TCR-α deficiency	/
IL-10-/-**	TNF$^{\Delta ARE*/+}$	Oxazalone	
Tg$_\varepsilon$26**		WASP deficiency	
Gi2α-deficiency*			
T-bet Tg			
STAT4 Tg*			
DSS*			
IL-7 Tg*			
MDR1α-deficiency			
TGFβ RII negativ Tg**			

Table 1 shows a sample of murine IBD models. Most IBD models develop a Th1-mediated experimental inflammation in the colon. The two exisiting ileitis models also show a Th1-type experimental inflammation, mimicking CD. Interestingly, there is not a single experimental IBD model that exhibits Th2-mediated ileitis. Mouse models that are indicated with * are members of the group of Type I models (induced by increased effector functions) whereas ** indicates that the model belongs to the group of Type II models (due to disturbed regulatory mechanisms). Abbreviations: SCID; severe combined immuno-deficiency, DSS; dextran sodium suphate, Tg; transgene, Gi; G protein, MDR; multi drug resistance, TCR; T cell receptor, WASP; Wiskott-Aldrich-Syndrome Protein

TNF$^{\Delta ARE/+}$ mice are a notable exception as this type I IBD model develops mucosal inflammation in the terminal ileum due to constant TNF overproduction. The increase in TNF is due to a knockout of adenosine uracil-rich elements (ARE) in the untranslated regulatory region of the TNF gene, resulting in increased translation of TNF. The disease phenotype in TNF$^{\Delta ARE/+}$ mice is very similar to CD, showing transmural lesions and granulomatosis. Interestingly, chronic ileitis and arthritis seem to be the only phenotypical consequences of constant TNF-overproduction in these mice (Kontoyiannis et al., 1999).

In contrast to TNF$^{\Delta ARE/+}$ mice, IL10-/- mice constitute a classical type II IBD model as mucosal inflammation in the large intestine is caused by the knockout of IL-10, an important regulatory cytokine. Lack of IL-10 results in defective immuno suppressive mechanisms and consequently, IL-10-/- mice develop a Th1-type colitis showing epithelial cell hyperplasia, abcesses and transmural inflammation. Most important, IL10-/- mice, like most experimental

IBD models, do not develop chronic inflammation when they are raised under germfree conditions (Strober et al., 2002). This fact again underlines the major impact of the intestinal microbiota on the initiation and development of IBD and sets the stage for IBD therapies aiming at changing the intestinal microbiota towards a more "regulatory" one. In this context, probiotic therapy was found to be an effective tool.

1.3 Probiotics: Clinical relevance and molecular mechanisms

1.3.1 History and definition of probiotics.

The term "probiotics" originates from the greek expression "pro bios" (for life). Fermentation of milk products by lactic acid bacteria to prevent food rotting was used in many cultures for thousands of years without even knowing about the existence of microorganisms. The direct positive effect of live bacteria within these products was recognized in the 19th century and this knowledge was subsequently used to treat diarrhea. Elie Metchnikoff is regarded as the father of the modern probiotic theory as he was the first to postulate the hypothesis that the uptake of beneficial lactic acid bacteria might be protective by replacing harmful bacteria in the intestine in 1907. In the 1950s, a preparation of *Bacillus subtilis*, Bactisubtil[R], was developed in France to restore the intestinal microbiota after antibiotic treatment (Schmoeger, 1960). Bactisubtil[R] thereby can be seen as the first commercially available probiotic product. The definition of probiotics changed several times ranging from `growth promoting substances produced by microorganisms` (Lilly and Stillwell, 1965) and `living microorganisms as food supplements, which positively influence the intestinal homeostasis of animals` (Fuller, 1989) to the definition from the WHO/FAO in 2001, which defined probiotics as microorganisms that confer a health benefit to the host, if they are taken up in sufficient amounts.

According to the WHO/FAO definition, the currently sold "probiotic" functional food products do not contain probiotics, as there are no placebo controlled studies proving the claimed general benefits for the intestinal immune system of healthy consumers. The only study reporting beneficial effects of a specific probiotic strain on healthy individuals showed that the uptake of *L. reuteri* increases work place healthiness (Tubelius et al., 2005). However, specific probiotic strains were convincingly proven to have protective activites in the prevention and therapy of very specific intestinal disease indications. Great success of specific probiotic strains was reported in the prevention and treatment of antibiotic-associated diarrhea or travellers diarrhea (Sazawal et al., 2006) as well as in the prevention of necrotising enterocolitis in preterm infants (Alfaleh and Bassler, 2008). Taking into account that dysregulated bacteria-host interaction is thought to be the underlying cause for the development and progression of IBD, the uptake of probiotics was hypothesized to be a sensible therapeutical approach.

1.3.2 Clinical relevance of probiotics in the context of IBD.

Up to now, IBD can not be cured and all therapeutical approaches aim to suppress disease symptoms like abdominal pain, bleeding and diarrhea in the acute phase or to prolong remission. The status quo in IBD treatment is the use of anti-inflammatory non-steroidal drugs like mesalazine, steroid-analoga like cortisol, anti-TNF antibodies and/or immuno-suppressive agents. Lack of response towards medication results in a significant number of patients who have to undergo surgical resection (Hwang and Varma, 2008) with a high risk of post-operative disease recurrence. Apart from that, all current treatment approaches are hampered by severe side-effects. In this context, probiotics were thought to be a safe and promising therapeutic alternative due to their microbiota-modulating activities. In consequence, the protective potential of many different probiotic preparations was investigated in numerous clinical studies. Unfortunately, many of these clinical studies were performed unblinded and are hampered by low numbers of study participants and/or the lack of adequate controls, making it impossible to draw firm conclusions. VSL#3, a mixture of eight bacterial strains and the single bacterial strain *Escherichia coli* Nissle 1917 (*E. coli* Nissle) are the only probiotics that were repeatedly found to have protective effects on IBD in well-designed clinical studies (Table 2).

    E. coli Nissle was found to be equivalent to mesalazine standard treatment in the maintenance of remission in UC patients (Kruis et al., 2004; Kruis et al., 1997; Rembacken et al., 1999). VSL#3 has been shown to prevent the onset of post-operative inflammation of the ileal pouch in UC patients, the so called pouchitis (Gionchetti et al., 2003; Pronio et al., 2008). Furthermore, VSL#3 was revealed to be an effective mean to maintain remission in pouchitis patients (Gionchetti et al., 2000; Mimura et al., 2004; Venturi et al., 1999).

First small and uncontrolled studies suggested VSL#3 to be effective in the maintenance of remission in UC patients (Venturi et al., 1999) and in the reduction of active UC or pouchitis (Bibiloni et al., 2005; Gionchetti et al., 2007). Additional open label studies revealed better results for the induction of remission in mildly to moderately inflamed UC patients by VSL#3 in combination with low balsalazide treatment compared to mesalazine or balsalazide treatment alone (Tursi et al., 2004). Nevertheless, well-designed studies with higher numbers of study participants are necessary to prove these preliminary results. In contrast to the promising results of probiotic therapy in pouchitis and UC patients, CD was found to be unresponsive towards probiotic therapy in several studies.

Table 2. Probiotic efficacy in clinical studies.

Disease	Publication	Probiotic Compound	Study Design	Outcome
Pouchitis	(Pronio et al., 2008)	VSL#3	PC, 31	Effective prevention
	(Mimura et al., 2004)	VSL#3	R, PC, 36	Maintenance of remission
	(Gionchetti et al., 2003)	VSL#3	R, PC, 40	Effective prevention
	(Gionchetti et al., 2000)	VSL#3	R, PC, 40	Maintenance of remission
	(Kuisma et al., 2003)	LGG	R, PC, 20	No reduction of active disease
Ulcerative colitis	(Furrie et al., 2005)	*B. longum*/Synergy	R,PC, 18	No reduction of active disease
	(Cui et al., 2004)	Bifidobacteria, BIFICO	R, PC, 30	Maintenance of remission
	(Kruis et al., 2004)	*E. coli* Nissle	R, 5-ASA, 222	Equivalent in maintenance of remission
	(Kato et al., 2004)	*Bifidobacterium breve/bifidum, Lactobacillus acidophilus*-fermented milk	R, PC, 20	Induction of remission
	(Ishikawa et al., 2003)	Bifidobacteria	R, PC, 21	Maintenance of remission
	(Rembacken et al., 1999)	*E. coli* Nissle	R, 5-ASA, 120	Equivalent in maintenance of remission
Crohn's Disease	(Chermesh et al., 2007)	Synbiotic 2000	R, PC, 30	No prolongation of remission
	(Van Gossum et al., 2007)	LA1	R,PC, 70	No prolongation of remission
	(Marteau et al., 2006)	LA1	R,PC, 98	No prolongation of remission
	(Bousvaros et al., 2005)	LGG	R, PC, 75	No prolongation of remission
	(Schultz et al., 2004b)	LGG	R,PC, 11	No reduction of active disease
	(Prantera et al., 2002)	LGG	R, PC, 45	No prolongation of remission
	(Malchow, 1997)	*E. coli* Nissle	R, PC, 75	No prolongation of remission

Table 2 summarizes published clinical studies investigating the effects of probiotics on IBD in a controlled and blind manner. Abbreviations: R; randomized, PC; placebo controlled, 5-ASA; mesalazine, n°; number of study participants, LGG; *Lactobacillus rhamnosus* GG, LA1; *Lactobacillus johnsonii* LA1

In summary, probiotics were shown to be effective in the prevention of pouchitis as well as in the maintenance of remission in UC and pouchitis patients whereas there is only preliminary data indicating probiotic efficacy in the reduction of active disease or CD. It became clear that the efficacy of probiotic therapy greatly varies dependent on the probiotic strain and the respective IBD indication. The clinical outcomes suggest that probiotic efficacy might depend on the induction of protective mechanisms that are no longer intact or overwritten in highly active inflammation, indicating that the right timing of probiotic supplementation plays a considerable role for probiotic efficacy. Furthermore, the protective mechanisms induced by probiotics seem to be disease and/or intestinal segment specific, as probiotic therapy was found to be uneffective in CD, a disease which differs from UC by being a typical Th1-type disease which does often affect the ileum. The only study indirectly reporting protective activity of a certain bacterial strain in the context of CD revealed that the amount of the commensal *Faecalibacterium prausnitzii* attached to the ileal mucosa inversely correlated with the rate of postoperative relapse (Sokol et al., 2008). To date, no study addressed the question of whether the lack of significant probiotic effects on CD might be due to the partly ileal disease location. In consequence, there is a clear need for large, well-designed clinical studies in order to draw reliable conclusions for the efficacy of every single probiotic preparation in the context of every single IBD indication.

1.3.3 Probiotic treatment results in increased IEC barrier functions and normalisation of the intestinal immune response.

Although probiotic efficacy in the prevention and treatment of several IBD indications has been proven in clinical studies, the molecular mechanisms underlying the protective effects are unclear. In order to shed some light on these mechanisms a lot of *in vitro* as well as experimental studies in IBD animal models addressed this question (Table 3).

Probiotics were found to protect intestinal epithelial barrier function by exerting anti-apoptotic effects on IEC. Salmonella-induced reduction of transepithelial resistance as well as increase of IEC apoptosis were found to be counteracted by secreted proteins of VSL#3 (Madsen et al., 2001). The anti-apoptotic effect of these probiotic proteins was found to be dependent on the induction of protective heat shock proteins (hsp) in IEC (Petrof et al.,

2004), but the bacterial proteins as well as their target receptors on IEC remained unidentified. The inhibition of TNF-induced pro-apoptotic signalling in IEC by LGG was found to be mediated by two secreted proteins, p40 and p75, which induced Akt-protein kinase dependent survival pathways via the activation of the epidermal growth factor receptor (Yan et al., 2007).

The reduction of active inflammation in UC patients by Bifidobacteria-fermented milk was correlated with increased luminal levels of butyrate, proprionate and other SCFA. This finding indicates that increased IEC nutrient supply by probiotics might be protective through mechanisms that counteract the loss of energy homeostasis and the subsequent induction of apoptosis in IEC (Kato et al., 2004).

In addition to protective effects on IEC viability, increased paracellular leakage due to reduced and dysregulated TJ proteins in IBD patients was revealed to be counteracted by probiotics. E. coli Nissle was found to inhibit enteropathogenic E. coli-mediated redistribution of the TJ protein zonula occludens (ZO)-2 in IEC in vitro (Zyrek et al., 2007). In addition, monocolonization of gnotobiotic mice with E. coli Nissle was found to increase the expression of the TJ protein ZO-1 and to reduce dextran sodium sulphate (DSS)-induced colitis in these mice (Ukena et al., 2007).

Apart from its effects on enterocytes, VSL#3 was found to increase intestinal mucin secretion by Goblet cells in healthy Wistar rats. The observed effect of the probiotic mixture was found to be induced by a heat, protease and DNAse insensitive compound that was mainly secreted by the lactobacilli group of the probiotic mixture (Caballero-Franco et al., 2007). In contrast, DSS-induced extensive disruption of the mucus and the IEC layer, resulting in subsequent development of severe experimental colitis, was not prevented or reduced by VSL#3 treatment (Gaudier et al., 2005). However, additional studies in other experimental models of IBD are necessary to answer the question of whether VSL#3-induced increased mucin secretion contributes to the protective effects of the clinically relevant probiotic mixture VSL#3.

In addition to protective effects on the physical epithelial barrier, probiotics were also shown to strengthen the first line of defense against luminal contents that is set up by IEC. Probiotics were found to increase IgA secretion in healthy mice as well as in DSS colitis, resulting in amelioration of DSS-induced inflammation (Vetrano et al., 2008). Furthermore, the flagellum of E. coli Nissle as well as unknown components of VSL#3 were found to induce the production of ß-defensins via the activation of NFκB and activating protein (AP)-1 dependent pathways in IEC in vitro (Schlee et al., 2008; Schlee et al., 2007; Wehkamp et al., 2004).

In addition to protective effects on IEC barrier function, probiotic bacteria were found to have suppressive effects on increased IEC activation, resulting in reduced secretion of proinflammatory chemokines. Secreted factors of *E. coli* Nissle were shown to inhibit TNF-induced secretion of the neutrophil attracting chemokine IL-8 without inhibiting NFκB activation in IEC *in vitro* (Kamada et al., 2008). Interestingly, VSL#3-derived DNA was also reported to reduce expression of IL-8 in IEC *in vitro*. In contrast to the effect of *E. coli* Nissle, the inhibition of IL-8 by VSL#3 was mediated by the reduction of proteasomal activity in IEC, resulting in blocked degradation of IκB, a major inhibitor of NFκB in IEC (Jijon et al., 2004). Consistently, soluble proteins of the probiotic mixture were found to inhibit *Salmonella dublin*-induced activation of NFκB and subsequent secretion of MCP-1 via proteasomal inhibition in IEC *in vitro* (Petrof et al., 2004). Although probiotics are convincingly reported to reduce IEC activation as well as chemokine secretion in cell culture systems, studies in experimental animal models are lacking in order to prove the physiological relevance of these mechanisms. In addition, although IP-10 was found to be a major proinflammatory chemokine in IBD (Banks, 2003; Hyun et al., 2005; Suzuki, 2007), no study addressed the question whether probiotics are able to modulate the expression of this specific chemokine in IEC. In addition to increased recruitment and activity of immune effector cells by chronically activated IEC, intrinsic dysregulation of immune effector cells plays a major role in IBD. In this context, probiotics were found to exert regulatory effects on adaptive immune responses.

A clinical study in pouchitis patients revealed that oral uptake of VSL#3 induced the proliferation of mucosal CD4$^+$CD25$^+$ and CD4$^+$/latency-associated peptide (LAP)$^+$ regulatory T cells, which are known to confer immunosuppressive effects. In addition, VSL#3 was found to increase the transcription of regulatory Foxp3 whereas the expression of proinflammatory IL1β was reduced in the mucosal tissue of these patients (Pronio et al., 2008). In consistence with clinical data, the induction of tolerance and immunosuppressive mechanisms by VSL#3 was proven in several animal models of chronic colitis. The probiotic mixture increased the expression of IL-10 and induced the proliferation of TGFβ bearing regulatory T cells resulting in reduced trinitrobenzene sulfonic acid (TNBS)-induced colitis (Di Giacinto et al., 2005). A recent study showed that probiotic-reduced TNBS-colitis is correlated with increased numbers of CD4$^+$CD25$^+$ as well as γδ-intraepithelial lymphocytes (IEL), which are thought to exert regulatory activites (Roselli et al., 2009). Whether the observed induction of regulatory T cells by VSL#3 in IBD and experimental inflammation is due to a direct effect of VSL#3 on mucosal T cells or due to regulatory effects of probiotics on APC or IEC is unknown.

A retrospective analysis of UC patients that were treated with VSL#3 revealed reduced expression of IL-8, IFNγ and IL1β as well as reduced numbers of polymorphnuclear cells in mucosal tissues (Lammers et al., 2005). Furthermore, biopsies of pouchitis patients treated with VSL#3 revealed reduced levels of TNF and IL1α whereas the expression of antiinflammatory IL-10 was increased. In addition, the authors reported reduced activity of tissue damaging enzymes such as gelatinase and nitric oxid synthase in the mucosa of probiotic-treated patients (Ulisse et al., 2001). VSL#3 and VSL#3-derived DNA were found to significantly reduce histopathological inflammation in IL10$^{-/-}$ mice. This effect was correlated with reduced mucosal and splenic expression of the proinflammatory cytokine IFNγ, indicating the reduction of Th1 cell activity by VSL#3 (Jijon et al., 2004; Madsen et al., 2001). However, it needs to be elucidated which effects are directly induced by probiotic bacteria and which ones are secondary effects due to reduced inflammation. With regard to this question, it is pivotal to reveal those host cells that are directly affected by probiotic activity in IBD animal models.

In this context, the bifidobacteria subgroup of VSL#3 was shown to exhibit potent anti-inflammatory activities through mechanisms that regulate antigen presentation and cytokine expression of dendritic cells (DC), suggesting that probiotics are capable to modulate the generation of T cell subsets. Intestinal- and blood derived dendritic cells produced high levels of IL-10 after stimulation with cell wall components of VSL#3-derived bifidobacteria whereas the expression of pro-inflammatory IL-12 and the costimulatory molecule CD80 was significantly reduced, indicating impaired capability of these DC to induce the maturation of Th1 cells (Hart et al., 2004). Similarly, Hoarau et al. showed that a fermentation product of *Bifidobacterium (B.) breve* induced maturation, enhanced IL-10 expression and prolonged survival of DC in a phosphoinositol-3-kinase-dependent manner *in vitro* (Hoarau et al., 2008). Most important, the transfer of *Lactobacillus rhamnosus* Lr32-treated bone marrow derived dendritic cells (BMDC) was shown to ameliorate TNBS-induced experimental colitis. In this study, the observed regulatory activity of probiotic-treated dendritic cells was found to be dependent on intact TLR2 and NOD signaling as well as on the induction of regulatory CD4^{+}CD25^{+} T cells via IL-10 independent mechanisms (Foligne et al., 2007).

Table 3. Probiotic effects on experimental inflammation in IBD animal models.

Animal Model	Publication	Probiotic Compound	Bacterial Component	Probiotic effect associated with reduced colitis
IL10-/-	(Jijon et al., 2004)	VSL#3	DNA	Reduction of proinflam. cytokines/chemokines systemic effect
	(Madsen et al., 2001)	VSL#3	-	Enhanced Barrier
	(Kamada et al., 2005)	E. coli Nissle	Heat-killed bacteria, bacterial DNA	-
SCID transfer	(van der Kleij et al., 2008)	L. reuteri, B. infantis	-	Reduction of proinflamm. cytokines/enzymes
	(Schultz et al., 2004a)	E. coli Nissle	-	Reduced MLN-derived proinflammatory cytokines
DSS	(Ukena et al., 2007)	E. coli Nissle	Cell surface structures	Enhanced barrier
	(Vetrano et al., 2008)	Colifagina lysed bacteria	-	Increased IgA secretion
	(Kamada et al., 2005)	E. coli Nissle	Heat-killed bacteria, bacterial DNA	-
	(Foligne et al., 2007)	LGG-stimulated BMDC	Cell wall components	Induction of regulatory DC
	(Rachmilewitz et al., 2004)	VSL#3/E. coli oral/sc	Unmethylated DNA	Systemic effect
	(Fitzpatrick et al., 2007)	VSL#3	-	Reduction of inflammatory cytokines/enzymes
	(Kokesova et al., 2006)	E. coli Nissle, L. casei, E. coli	-	-
	(van der Kleij et al., 2008)	L. reuteri, B.infantis	-	Reduction of inflammatory cytokines/enzymes
TNBS	(Di Giacinto et al., 2005)	VSL#3	-	Induction of regulatory T cells
	(Roselli et al., 2009)	(L. acidophilus, B. longum), (L. plantarum, S. thermophilus, B. animalis subsp. lactis)	-	Induction of regulatory intraepithelial lymphocytes Reduced proinflammatory cytokines/chemokines
	(Foligne et al., 2007)	L. rhamnosus LR32 - stimulated BM-DC	Cell wall components	Induction of regulatory DC
	(Peran et al., 2006)	L. fermentum	-	Reduction of oxidative stress
Iod-acetamid	(Shibolet et al., 2002)	VSL#3/LGG	-	Reduction of inflammatory cytokines/enzymes

Table 3 summarizes all experimental studies that showed significant protection of probiotic therapy in the context of experimental colitis. Reduced inflammation was found to be due to various probiotic effects on the level of intestinal barrier function and IEC, APC or immune effector cell function. All studies reporting protective effects of probiotics were conducted with colitis models whereas there is no data about probiotic effects on experimental ileal inflammation. Abbreviations: SCID; severe combined immunodeficiency; MLN; mesenteric lymph nodes

In summary, it was found that probiotics are able to counteract an array of intestinal disturbances that are known to contribute to the development and progression of IBD. However, from a mechanistical point of view, probiotic efficacy is still far from being understood. Several *in vitro* studies revealed barrier protective and anti-inflammatory mechanisms of probiotics in IEC or DC but almost none of these mechanisms were subsequently tested with regard to their physiological relevance. On the other hand, experimental studies showing protective effects of probiotics on experimental colitis mostly remain on a descriptive level. In most studies, it is completely unclear which cells were the primary probiotic target cells and which molecular mechanisms initiated the anti-inflammatory cascade. In addition, studies addressing the efficacy of probiotics in ileal inflammation are lacking. However, many studies showed that live bacteria are not always necessary for the induction of protective mechanisms. The finding that cell-wall associated components, secreted proteins or bacterial DNA can be equally effective than live bacteria revolutionizes the probiotic theory once again and is even contradictory to the current definition of probiotics. Nevertheless, the identification and isolation of these active bacterial structures would open the opportunity to develop new therapeutical approaches in the context of IBD.

2   AIMS OF THE WORK

Whereas many clinical and experimental studies revealed protective effects of probiotics in the context of IBD, the underlying protective mechanisms are unclear. There is almost nothing known about bacteria-host interactions with regard to active probiotic structures, host target cells and receptors as well as signalling pathways that are induced in host cells. Consistent with the lack of mechanistical understanding, it is completely unknown why probiotic therapy was found to be effective in the prevention and treatment of UC and pouchitis, but not in CD. The intestinal epithelium, which constitutes the borderline between luminal bacteria and host tissue, was shown to have important regulatory functions on the intestinal homeostasis, suggesting that probiotic effects on this particular cell type can be of high relevance in the treatment and prevention of IBD.

Hence, the present work aimed to unravel structure-related anti-inflammatory mechanisms of VSL#3, a clinically protective probiotic mixture, in IEC. A major interest of the present study was to subsequently analyse the physiological relevance of these probiotic mechanisms. In order to contribute to the understanding of why probiotic therapy is protective in some IBD indications but not in others, the anti-inflammatory effect of VSL#3 on intestinal compartment specific inflammation was investigated in two murine experimental models for IBD, exhibiting ileal or cecal/colonic experimental inflammation.

3   MATERIAL AND METHODS

Cell culture. The small intestinal epithelial cell line Mode-K (Vidal et al., 1993) (passage 10-25), the human embryonic kidney epithelial cell line (HEK) 293 (passage 5-25), TLR2 deficient HEK cells (Cat.nr.:293-null) and TLR2-positive HEK cells (Cat.nr.:293-mtlr2) (Invitrogen, Carlsbad, USA) (passage 2-5) as well as T84 cells (differentiating, polarizing human colon carcinoma cells) were grown in a humidified 5% CO_2 atmosphere at 37°C. Cells were grown to confluency in 6, 12 or 24 well tissue culture plates (Greiner bio-one, Frickenhausen, Germany) prior to stimulation experiments. Mode-K cells were grown in Dulbecco´s modified Eagle´s medium (DMEM) (Invitrogen, Carlsbad, USA) containing 10% fetal calf serum (FCS) (Invitrogen, Carlsbad, USA), 1,0% Glutamine (Invitrogen, Carlsbad, USA) and 0,8% antibiotic antimycotic (AA) (Invitrogen, Carlsbad, USA). Glutamine was omitted for the cultivation of HEK293 cells and Blasticidin (10 µg/ml) (InvivoGen, San Diego, USA) was added to HEK cell culture media for the cultivation of stably transfected HEK293 cells. Cell culture media was changed prior to cell culture stimulation experiments. T84 cells were cultivated in DMEM/Ham F12 (Invitrogen, Carlsbad, USA) (10% FCS, 0,8% AA).

Bacterial culture and treatment. Lyophilized VSL#3 bacteria (a generous gift from Dr. DeSimone, L´Aquila, Italy) were resuspended in DMEM before they were used in cell culture stimulation experiments. VSL#3 derived *L. casei* (reclassified as *L. paracasei* according to VSL#3 pharma in 2008) (a generous gift from Dr. DeSimone, L´Aquila, Italy), *L. casei* BL23, *L. plantarum* 299v (generous gifts from Dr. Gaspar, Instituto de Agroquimica y Tecnologia de Alimentos, Valencia, Spain) and *L. casei* ATCC393 were grown at 37°C in MRS Broth (Fluka, Heidelberg) containing 0.05% L-cysteine (Roth) under anaerobic conditions using Anaerogen packages (Anaerogen, Oxoid, UK). *E. coli* strain Nissle 1917 (a generous gift from Dr. Sonnenborn, Ardeypharm GmbH, Herdecke, Germany) was grown aerobically in LB-medium (AppliChem, Darmstadt, Germany). Bacteria were centrifuged (4500 g, 10 min), washed in an equal volume of PBS (1x) (4500 g, 10 min) and resuspended in DMEM. To characterize effective probiotic components, *L. casei* was heat-killed (90°C, 30 min), fixed (fl.c) (5% formaldehyde, 3 h, 4°C) or lysed by lysozyme (50 mg/ml) (Sigma, Steinheim, Germany) in filter sterilized Tris buffer (10 mM Tris, pH8). Fixed bacteria were washed three times with sterile 1xPBS (Invitrogen, Carlsbad, USA) and resuspended in the respective cell culture media. All cell culture experiments were performed with fixed bacteria, if not otherwise indicated. For cell surface treatment, bacteria were incubated with Phospholipase A (2 mg/ml) (Sigma, Steinheim, Germany), Trypsin (2 mg/ml) (Roth, Karlsruhe, Deutschland),

Proteinase K (50 µg/ml) (Roth, Karlsruhe, Deutschland) in filter sterilized Tris buffer (50 mM Tris, 0,1 M NaCl, pH 8) (37°C, 1 h) in a shaker.

Preparation of bacterial conditioned media (CM). CM was generated by growing VSL#3-derived *L. casei*, *L. casei* BL23, *L. casei* ATCC393 and *L. plantarum* 299v (5×10^7 cfu/ml) anaerobically in DMEM (1% Glutamine) containing HEPES (20 mM) (Invitrogen, Carlsbad, USA) (37°C, 5% CO_2, 24 h). CM was cleared from bacterial cells (4500 rpm, 10 min), titrated to a pH of 7,6 and filter-sterilized. Sterile CM was either supplemented with FCS and used to replace the cell culture media in cell culture stimulation experiments or concentrated using 10 or 50 kDa amicon filter devices (Millipore, Carrigtwohill, Ireland) and diluted to 1xCM in the cell culture supernatant. Where indicated, CM was heat-inactivated (60°C, 20 min), supplemented with PMSF (1 mM) (RT, 15 min) or underwent precipitation with 50% ice cold ammoniumsulfate (-20°C, o.n.) followed by resuspension of the dried precipitate (12000g, 4°C, 10 min) in DMEM.

Protease Assay. Serin/cystein protease activity of CM *L. casei* (VSL#3) and CM *L. casei* BL23 was analysed using a Protease Assay Kit (Cat. Nr. 539125, Calbiochem, Darmstadt, Germany) which is based on proteolytic degradation of FTC-casein according to the manufacturers instructions.

Cell culture stimulation. Confluent Mode-K or HEK cell monolayers were stimulated with TNF (10 ng/ml) (R&D Europe, Abington, England), IFNγ (50 ng/ml), brefeldin A (0.5 µM) (Calbiochem, Darmstadt, Deutschland), lactacystin (22 mM) (Biomol, Hamburg, Germany), NH_4Cl (20 mM) (Sigma, Steinheim, Germany), 3-methyladenine (3-MA) (5 mM) (Sigma, Steinheim, Germany), VSL#3, VSL#3-derived *L. casei*, *L. plantarum* 299 v, *L. casei* BL23, *L. casei* ATCC 393 (type strain) or *E. coli* Nissle 1917 (24 h, moi 20, if not otherwise indicated).

ELISA. IP-10 (murine/human) and IL-6 (murine) concentrations in IEC supernatants were determined using the appropriate ELISA kits (R&D Europe, Abington, England) according to the manufacturers instructions. Serum amyloid A (SAA) concentrations in plasma were determined using the appropriate ELISA kit (Invitrogen, Karlsruhe, Germany) according to the manufacturers instructions.

Chromatin Immunoprecipitation (ChIP). Mode-K cells in 75 cm^2 flasks (Greiner bio-one, Frickenhausen, Germany) were preincubated with *L. casei* (1 h) and stimulated with TNF (10 ng/ml) (2 h). Cells were fixed in formaldehyde fixing solution (1%) followed by nuclear extraction as well as chromatin immunoprecipitation (ChIP) analysis, using a ChIP-kit (Active Motif, Carlsbad, USA). Immunoprecipitation was performed using an anti-NFκB p65 antibody (Cell Signaling, Beverly, USA) (4°C, o.n.). DNA/protein/antibody-complexes were incubated with salmon sperm saturated protein A/G agarose (Santa Cruz, Heidelberg, Germany) (30 min) and washed in high salt buffer (3x) followed by washes in no salt buffer (3x). DNA was released from the immune complex by heating and subsequent proteinase K treatment. DNA was extracted using phenol-chloroform and eluted in water. Total DNA from the nuclear extract was used as input control for the PCR analysis. PCR was performed with 1 µl of DNA using the following IP-10 promoter-specific primers:

5´-AACAGCTCACGCTTTG, 5´-GTCCTGATTGGCTGACT.

The length of the amplified product was 186 bp. PCR products (10 µl) were subjected to electrophoresis on 2% agarose gels and visualized by Geneflash Imager (VWR, Ismaning, Germany).

Transfection. Mode-K cells (luciferase assay) or HEK293 cells (IP-10 overexpression assay) (50% confluent, 24-well plate) were transfected using FuGENE (Roche, Mannheim, Germany) according to the manufacturers instructions. A mixture of 23 µl DMEM, 1.5 µl FuGENE (Roche, Indianapolis, USA) and 0.5 µg of the appropriate plasmid (pGL3-basic-IP10 (p-IP-10), pGL3-basic (control-vector), pIP-10-DsRed, pDsRed-Monomer-C1 (control-vector) was added to every well and incubated for 24h (luciferase assay) or 6h (IP-10 overexpression assay) prior to stimulation experiments. Plasmids were (Hormannsperger et al., 2009).

Luciferase assay. Mode-K cells transfected with the luciferase-reporter plasmid or the control plasmid were stimulated with TNF or TNF and *L. casei* for 24 h. Cells were lysed in 30 µl lysis buffer (Promega, Madison, USA) and cell debris were separated by centrifugation (12000 rpm, 5 min). Supernatants (25 µl) were mixed with Reagent A (PJK, Kleinblittersdorf, Germany) and firefly luminescence was measured (550 nm). A volume of 100 µl of Reagent B (PJK, Kleinblittersdorf, Germany) was added and renilla luminescence was measured (480 nm). Relative luciferase activity was calculated in percentage: $((\text{firefly/renilla})_{sample}/(\text{firefly/renilla})_{control}) \times 100$.

RNA isolation, reverse transcription and real-time PCR. RNA isolated from Mode-K cells or primary IECs was extracted using Trizol Reagent (Invitrogen, Karlsruhe, Germany) according to the manufacturers instructions. Extracted RNA was solved in 20 µl DEPC-water (0.1%). RNA concentration and purity (A_{260}/A_{280} ratio) was determined by spectrophotometric analysis (ND-1000 spectrophotometer, NanoDrop Technologies, Willigton, USA). Reverse transcription was performed of 1 µg total RNA using M-MLV (Invitrogen, Karlsruhe, Germany). Real-time PCR was performed using 1 µl cDNA in a Light Cycler™ system (Roche Diagnostics, Mannheim, Germany) as previously described (Ruiz, 2005). Primer sequences were the following:

TLR2	Sense	5'-TGGGGGTAACATCGCT
	Reverse 5'-CATCTACGGGCAGTGG	
IP-10	Sense	5'-TCCCTCTCGCAAGGAC
	Reverse 5'-TTGGCTAAACGCTTTCAT	
18S	Sense	5'-CGGCTACCACAT-CCAAGGAA
	Reverse 5'-GCTGGAATTACCGCGGCT	

The amplified product was detected by the presence of a SYBR green fluorescent signal. Melting curve analysis was used to document amplicon specificity and crossing points (Cp) were determined. Relative induction of mRNA expression was calculated according to the $2^{-\Delta\Delta Cp}$ method (Pfaffl, 2001) and normalized to the expression of 18S. Data were expressed as fold change versa the indicated control cells.

Presence or absence of TLR2 transcripts in the transfected HEK293 cells was visualized by running a 2% Agarose gel with real time PCR amplicons (321 bp) generated after reverse transcription of mRNA.

T cell transmigration assay. Murine splenocytes were isolated from a fresh spleen (mouse genotype: C57Bl//N) and activated with 2.5 µg/ml Concavalin A (Sigma, Steinheim, Germany) in RPMI (Invitrogen, Karlsruhe, Germany) (10% FCS, 37°C, 5% CO_2, 16 h). Cells were then centrifuged and 1×10^8 cells were resuspended in 1 ml RPMI (25 mM Hepes). The cell suspension was given onto a 5 ml NycoPrep 1.077 A (AXIS-SHIELD, Oslo via Progen) and centrifuged (600 g, 20 min). The activated T lymphoblasts were taken and resuspended in 30 ml transmigration media (RPMI, 5% FCS, 25mM Hepes). The cells were washed three times (250 g, 10 min) and resuspended to a final cell number of 2×10^6 cells/ml transmigration media. The activated murine T lymphoblasts were seeded on top of transwell filters (5 µm) at a concentration of 2×10^5 cells/ml transmigration media. The lower chamber was filled with 500 µl of transmigration medium and 100 µl of pure Mode-K cell culture media or cell culture media from stimulation experiments with Mode-K cells (control, TNF, TNF/fL.c) or TNF-conditioned media supplemented with either a neutralizing anti-IP-10 antibody (30μg/ml) (R&D Europe, Heidelberg, Germany) or a control goat-IgG (30 μg/ml) (Dianova, Hamburg, Germany). The assay was then incubated (37°C, 5% CO_2, 2 h) and the number of transmigrated cells in the lower chamber was analyzed using an Axiovision Cell Counter (Zeiss, Göttingen, Germany). The assay was performed in triplicates and the number of transmigrated cells at three representative regions (roi) of each single triplicate was measured.

Western Blot. Protein isolated from primary IEC or Mode-K cells that were taken up in Laemmli buffer (TrisCl (12,5 ml), 10% SDS (8 ml), Glycerol (7,9 g), DTT (7,7 mg), Bromphenole (0,5%) (1,25 ml), H_2O (50 ml)) (200 μl/sixwell after PBS washes (2x)) was denatured (95°C, 15 min). 25 μg of protein were subjected to sodium-dodecyl-sulfate polyacrylamide gel electrophoresis (SDS-PAGE) on 10%/15% SDS-PAGE gels. Anti IP-10 (R&D Europe, Heidelberg, Germany, Arlington), anti-IκB, anti-phospho-RelA^{Ser536}, anti-ubiquitine, anti-phospho-EIF4E, anti-phospho-EIF4G (Cell Signaling, Beverly, MA), anti-DsRed (clontech) and anti-ß-actin-antibody (ICN, Costa Mesa, CA) were used to detect immunoreactive IP-10, IκB, phospho-RelA, phospho-EIF4E, phospho-EIF4G, DsRed, ubiquitine and ß-actin, using an enhanced chemoluminescence light-detecting kit (Amersham, Freiburg, Germany).

Pulse-chase experiment. Mode-K cells (6-well plates) in DMEM (0.5% FCS) were stimulated with TNF during a 3h pulse period (S^{35} Methionine/Cysteine Labeling Mix, 25 μCi/ml) (Perkin Elmer, Waltham, USA). After the pulse period, the cells were washed with 1xPBS (3x) and either lysed in 200 μl of 1x lysis buffer (Cell Signaling, Beverly, USA) supplemented with PMSF (1 mM) or they underwent an additional chase period of 3h with or without stimulation with *L. casei*. After the chase period, cells were washed with 1xPBS (3x) and lysed as described above. IP-10 co-immunoprecipitation was performed as described and beads were resuspended and cooked (95°C, 10 min) in 20 μl of Laemmli buffer for subsequent gel electrophoresis on a 15% SDS gel. After drying the gel it was placed on a Kodak Storage Phosphor screen (Amersham bioscience, Freiburg, Germany) in a cassette (Amersham Bioscience, Freiburg, Germany) (o.n.). The Storage Phosphor screen was then scanned by a Typhoon TRIO+scanner (Amersham Bioscience, Freiburg, Germany).

Electrospray liquid chromatography tandem mass spectrometry (ESI-LC-MS/MS) preparation and analysis. Active and non-active chromatographic fractions of CM *L. casei* were subjected to SDS-gel electrophoresis in order to remove contaminants that could disturb differential LC-MS/MS analysis. Gels were fixed (2% acetic acid, 40% MeOH) (RT, 1 h) followed by Coomassie Blue staining (16% Coomassie, 64% ddH$_2$O, 20% MeOH) (RT, 2 h) (4°C, o.n.). Gels were destained by a short wash (5% acetic acid, 25% MeOH) followed by two washes in 25% EtOH (RT, 1 h) and one wash in bdH$_2$O (RT, 1h). Stained gel pieces were subjected to in-gel trypsin digestion using triethylammonium bicarbonate (TEAB), formic acid (FA) (both Sigma-Aldrich, Steinheim, Germany) and acetonitrile (ACN) (Rathburn Chemicals, Walkerburn, Scotland) according to the following procedures:

	Vol/µL	Buffer	Time (min)	Temp (C°)	Spin (1000 rpm, 1 min)	Kept compartment
1. Destain	100	50% 5mM TEAB/50%EtOH	120	55	Yes	Pellet
2. Destain	100	50% 5mM TEAB/50%EtOH	60	55	Yes	Pellet
3. Dehydration	100	Absolute EtOH	10	RT	Yes	Pellet
4. Wash	100	5mM TEAB	20	RT	Yes	Pellet
5. Dehydration	100	Absolute EtOH	10	RT	Yes	Pellet
6. Dehydration	100	Absolute EtOH	10	RT	Yes	Pellet
7. Trypsin addition on ice	25/12	10 ng/µL trypsin in 5mM TEAB	15	4	/	/
8. Trypsin removal		Spin out residual trypsin			Yes	Pellet
9. Digestion buffer addition	20	5mM TEAB	4h	37	/	/
10. Acidification	5	5% FA	0	RT	Yes	Solution
11. First Extraction	20	1% FA	30	RT	Yes	Solution
12. Second Extraction	20	1% FA	30	RT	Yes	Solution
13. Third Extraction	20	60%ACN 40% 0.1% FA	30	RT	Yes	Solution
14. Drying Step	30	ACN	15	RT	/	/
15. Drying Step	20	ACN	15	RT	Yes	Solution

Peptides were then lyophilized and stored at -20°C until they were resuspended in 20 µl 0.1% FA. ESI-LC-MS/MS analysis was performed on an amaZon ETD iontrap mass spectrometer (Bruker Daltonik GmbH, Germany) coupled to an easy nLC nano LC system (Proxeon Biosystems, Odense, Danmark). Buffer A contained 0.1% FA in ddH_2O and buffer B 0.1% FA in ACN. The injection volume was 20 µl per sample, and after precolumn equilibration (5 min, 2 µl/min, buffer A) and analytical column equilibration (5 min, 300 nl/min, buffer A) the sample was separated over a 110 min gradient going from 0% buffer B to 40% buffer B at a flow rate of 300 nl/min. The peptides eluting off the column were sprayed directly into the mass spectrometer where the masses of intact peptides were acquired in enhanced scan mode (8300 amu/s). Subsequent mass spectrometry analysis was performed with on-the-fly determination of precursor charge state, exclusion of singly charged ions and further fragmentation of the five most intense precursor ions.

Protein identification was performed using Mascot 2.2.04 and ProteinScape 2.0-MR 42 searching for sequence hits in Swissprot (Bacteria-Firmicutes). Search parameters were chosen as follows: Carbamidomethylation and methionine oxidation were taken into account as variable modification, and a tolerance of 0.3 Da was applied for both, peptides and fragments. Proteins identified by at least one peptide showing a fragment score over 30 were regarded as hits.

Co-immunoprecipitation. Mode-K cells (6-well plates) were stimulated with TNF or TNF and *L. casei* (6 h). Cells were lysed in 200 µl of 1x lysis buffer (Cell Signaling, Beverly, MA) supplemented with PMSF (1 mM). Cell debris were removed by centrifugation (1400 g, 10 min) and supernatants were incubated with anti-IP-10-antibodies (R&D Europe, Heidelberg, Germany) (4°C, 3 h) in a shaker. Samples were then incubated (4°C, o.n.) together with 13 µl of protein A/G beads (Santa Cruz, Heidelberg, Germany), which were previously washed twice with 1x lysis buffer (Cell Signaling, Beverly, USA) in a shaker. Beads were collected by centrifugation (8000 g, 5 min), washed with 1x lysis buffer (2x) and resuspended in 50 µl Laemmli buffer for subsequent Western blot analysis.

Transwell experiments. T84 cells were grown to confluency in the apical compartment of transwells (0,4 µm) (Corning, NY, USA) according to the manufacturers instructions. When T84 cells had built up a stable transepithelial resistance (TEER)(Ohm/cm^2) measured by using a voltohmeter at 37°C (Millipore, Millicell, ERS), the respective stimulation experiments were performed in triplicates. IFNγ (50 ng/ml)/TNF (20 ng/ml) stimulation was used as a positive control for TEER reduction. The effect of the stimulants on TEER was measured at 30 min, 1 h, 2 h, 3 h, 6 h, 12 h and 24 h after the stimulation.

Animal models. Conventionally raised TNF$^{\Delta ARE/+}$ mice (a generous gift from Kollias G., Institute for Immunology, Biomedical Sciences Research Center "Al. Fleming", Greece) on C57BL/6 background as well as wildtype C57BL/6 were fed 1.3×10^9 cfu VSL#3 or VSL#3 derived *L. casei* in 13.2% (w/v) gelatine, 20% (w/v) glucose in water every weekday for 15 weeks. The gelatine was prepared freshly every third day. Probiotic feeding was started post-weaning at the age of three weeks. Placebo-fed mice were used as controls.

Analogously, TNF$^{\Delta ARE/+}$ and wildtype mice were fed VSL#3 from day one after birth for 18 weeks in a pre-weaning experimental setup with additional feeding of the mother mice (as soon as plaque was detected as a first indication of pregnancy until the end of the weaning period). The offspring was fed with liquid VSL#3-suspension that was prepared freshly every day for the first two weeks after birth. Placebo-fed mice were used as controls. Mice were killed at the age of 18 weeks.

SPF-raised IL-10-/- mice on a 129 SvEv/BL/6 background and 129 SvEv wildtype mice were fed VSL#3-gelatine every weekday for 21 weeks starting post-weaning at the age of three weeks. Placebo-fed mice were used as controls. Mice were killed at the age of 24 weeks. After sacrificing the mice, sampling of plasma and gut content as well as IEC isolation was performed. Sections of the distal ileum, cecal tip and distal colon were fixed in 10% neutral buffered formalin (Sigma Aldrich, Steinheim, Germany). Fixed tissues were hematoxylin-and eosin (H&E)-stained and embedded in paraffin. Histology scoring was performed by an independent pathologist in a blinded way, assessing the degree of lamina propria mononuclear cell infiltration, crypt hyperplasia, goblet cell depletion and architectural distortion, resulting in a score from 0 (not inflamed) to 12 (inflamed), as previously described (Katakura, 2005). Animal use was approved by the institution in charge (approval no. 55.2-1-S4-2531-74-06 and 32-2347/4+63).

Isolation of primary mouse IEC. Primary IEC were purified as previously described (Ruiz, 2005). Briefly, either ileal/jejunal or cecal/colonic tissue was cut into small pieces and incubated (37°C, 15 min) in Mode-K cell culture media supplemented with 1 mM DTT (Roth, Karlsruhe, Germany). The tissue/IEC suspensions were filtered, centrifuged (300 g, RT, 7 min) and cell pellets were resuspended in DMEM containing 5% FCS. The remaining tissue was incubated in 30 ml PBS (37°C, 10 min) containing 1.5 mM EDTA (Roth, Karlsruhe, Germany). After filtration, the tissue was discarded and the cell suspension from this step was centrifuged as above. Finally, primary IEC were loaded onto a 20%/40% discontinuous Percoll gradient (GE Healthcare, Uppsala, Sweden) and purified by centrifugation (600 g, 30 min). Primary IEC that assemble between the two phases were collected in proteome lysis buffer (7 M urea, 2 M thiourea, 2% CHAPS, 1% DTT) for subsequent protein isolation or in trizol (Qiagen, Maryland, USA) for subsequent RNA isolation. Purity of IEC was confirmed using anti-CD3$^+$-Western blot analysis.

DNA isolation from gut content and bacteria-specific PCR. DNA was extracted from 200 mg of gut content using the QIAamp DNA Stool Mini Kit (Qiagen, Maryland, USA) according to the manfacturers instructions. *Streptococcus thermophilus (S. thermophilus)* (Tilsala-Timisjärvi, 1997)- and *L. casei* (Haarman and Knol, 2006)-specific PCR of the rRNA intergenic spacer region was performed as previously described.

Sample preparation and molecular enumeration of mucosal *L. casei.* Distal ileal tissue segments (3 mm) were prepared according to published work (Conte MP, 2006). Briefly, segments were washed four times to prevent analysis of luminal bacteria. After preparation, bacterial and tissue pellets were obtained by centrifugation (8000 g, RT, 3 min). DNA was extracted from cell pellets using the DNeasy Blood & Tissue kit (Qiagen, Maryland, USA). *L. casei*-specific primer were used to perform real time PCR as previously described (Haarman and Knol, 2006). For calibration, mucosal tissue pieces of a non-treated wild-type mouse were spiked with 10-fold dilutions of *L. casei* (VSL#3), ranging from 5.4 × 10^3 to 10^6 cfu/ml.

Immunohistochemical labeling. Cecum was dissected and immediately immersed in Tissue Tek OCT compound (Agar Scientific, Stansted, UK) and frozen in liquid nitrogen. Samples were stored in liquid nitrogen. 8 µm sections were cut using a Leica CM1950 cryostat, picked up onto polylysine coated slides and air dried (RT, 45 min). Sections were fixed (4% paraformaldehyde, 0.1 M phosphate, pH 7.4, 5 min) before being washed with 6 changes of PBS (pH 7.4) over 20 minutes. Slides were then incubated with 0.1% Triton X 100 (Sigma, Steinheim, Germany) in PBS (3 min) before being washed as above. Fc receptors were blocked (Fcγ III/II receptor, BD Pharminogen Heidelberg, Germany) and slides incubated in 10% BSA 5% normal donkey serum in PBS (RT, 2 h). Blocking buffer was removed by capillary action and the slides incubated in either IP-10 goat polyclonal IgG (Santa Cruz G-15 sc-14641) or control goat IgG (Santa Cruz sc-2028) in 2% BSA (4°C, o.n.). The slides were washed with 6 changes of PBS over 1 hour. Alexa Fluor 488 donkey anti goat IgG (Invitrogen Molecular Probes) was applied to the sections and incubated (RT, 30 min). The slides were washed as above and mounted in DAPI-containing vectashield (Vector laboratories, Burlingham, USA). Sections were viewed on a Zeiss Axioskop microscope using a FITC and DAPI filter set and imaged using a QIMAGING camera.

Statistical analysis. Data were expressed as mean of triplicates +/- standard deviation. Statistical tests were performed using two-tailed Student test except for the *in vivo* feeding study as the normality test failed. In this case, rank sum test was applied. Differences were considered significant if values were < 0.05 (*) or < 0.01 (**). Experimental procedures were repeated in independent experiments as indicated.

4   RESULTS

4.1   *L. casei* (VSL#3) mediates functionally relevant selective inhibition of IP-10 secretion in IEC

4.1.1   *L. casei* (VSL#3) selectively inhibits IP-10 expression in IEC

VSL#3 was tested for effects on the expression of proinflammatory cytokines and chemokines in unstimulated and TNF-activated Mode-K cells. VSL#3 was found to significantly reduce the amount of TNF-induced IP-10 expression in IEC after 24 h of probiotic costimulation. In contrast, expression of the cytokine IL-6 was strongly induced by probiotic costimulation, suggesting a selective inhibitory mechanism of VSL#3 on IP-10 expression (Figure 6).

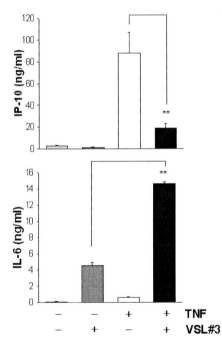

Figure 6: Probiotic effects on cytokine and chemokine secretion.
ELISA analysis showed that VSL#3 selectively reduces IP-10 concentration in the cell culture supernatant of TNF-activated Mode-K cells.

Stimulation experiments with the eight different bacterial strains of VSL#3 revealed that the specific inhibition of IP-10 by the probiotic mixture was mediated by a single one of these bacterial strains, by *L. casei* (VSL#3) (Figure 7). The seven other bacterial strains of VSL#3 did not exert an inhibitory effect on IP-10 expression (data not shown). *L. casei* (VSL#3) selectively inhibited TNF-induced IP-10 expression, whereas IL-6 expression (Figure 7) and the expression of another chemokine, macrophage inflammatory protein (MIP)-2 (data not shown), were increased by the probiotic strain.

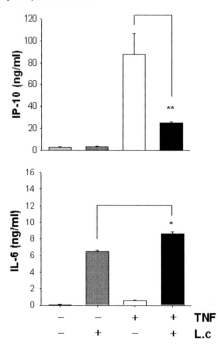

Figure 7: Probiotic effects on cytokine and chemokine secretion.
ELISA analysis shows that the stimulation of Mode K cells with living *L. casei* (VSL#3) is sufficient to induce selective inhibition of IP-10 in the cell culture supernatant of TNF-activated Mode K cells.

In order to assess the potential of other probiotic bacterial strains to induce similar effects as *L. casei* (VSL#3) in IEC, the effects of two more clinically effective probiotic strains were investigated. *E. coli* Nissle 1917 (E.c. Nissle) and *L. plantarum* 299v (L.p 299v), a probiotic that has been shown to be clinically effective in irritable bowel syndrome (Niedzielin et al., 2001), were used to stimulate TNF-activated Mode K cells. We found that the selective inhibition of IP-10 in IEC by *L. casei* (VSL#3) is not an ubiquituous characteristic of probiotic bacteria, given that *E. coli* Nissle and *L. plantarum* 299v did not exert analogous effects (Figure 8).

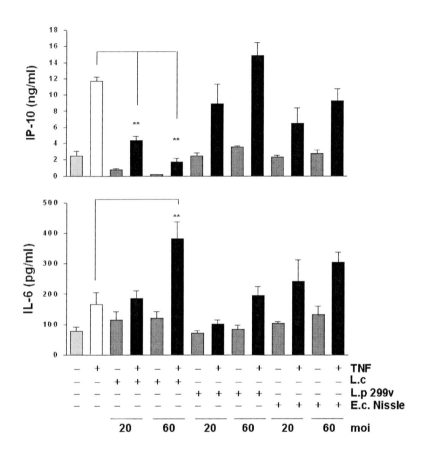

Figure 8: The selective inhibition of IP-10 secretion is a specific feature of *L. casei* (VSL#3).
ELISA analysis of cell culture supernatants after stimulation experiments with different probiotic bacterial strains in different mois revealed that only *L. casei* (VSL#3) is capable to selectively inhibit IP-10 secretion whereas *E. coli* Nissle (E.c. Nissle) and *L. plantarum* 299v (L.p 299v) do not have similar effects.

Interestingly, the inhibition of IP-10 was found to be mediated not only by *L. casei* (VSL#3) but also by *L. casei* ATCC393, the type strain of *L. casei*. In contrast, *L. casei* BL23, one of the two *L. casei* strains that are sequenced, did not exert similar effects (Figure 9). This finding suggests that the observed inhibition of IP-10 expression in IEC is a strain-specific effect.

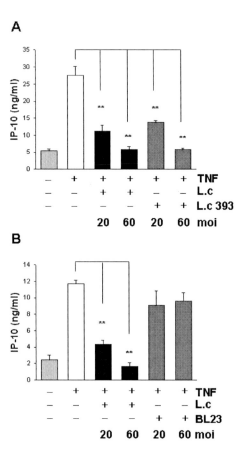

Figure 9: The type strain of _L. casei_ (_L. casei_ ATCC393) is equally potent to inhibit IP-10 secretion as _L. casei_ (VSL#3).

ELISA analysis of cell culture supernatants after stimulation experiments with _L. casei_ (VSL#3), _L. casei_ ATCC393 (A) and _L. casei_ BL23 (B) in different mois revealed that the _L. casei_ type strain is as potent to inhibit IP-10 secretion as _L. casei_ (VSL#3) whereas _L. casei_ BL23 has no effect on IP-10 secretion.

4.1.2 *L. casei* (VSL#3)-induced selective inhibition of IP-10 in IEC impairs pro-inflammatory T cell recruitment

T cell transmigration assays were performed in order to address the question whether the observed selective inhibition of IP-10 secretion in IEC by *L. casei* (VSL#3) is functionally relevant. Supernatants of TNF- and *L. casei* (VSL#3)-stimulated Mode K cells were analysed in a T cell transmigration assay using activated murine T lympoblasts. The culture supernatants of *L. casei* (VSL#3)-stimulated TNF-activated Mode K cells were found to induce significantly less migration of activated T lymphoblasts compared to culture supernatants derived from TNF-activated IEC without probiotic stimulation. In addition, neutralisation of IP-10 in the supernatant of TNF-activated IEC via anti-IP-10 antibodies resulted in strongly reduced numbers of transmigrated T cells (Figure 10), revealing that functionally active IP-10 is a major IEC chemokine in the context of T cell recruitment. These results suggest the observed selective inhibition of IP-10 in IEC by *L. casei* (VSL#3) to be of potential therapeutical relevance in the context of intestinal inflammation.

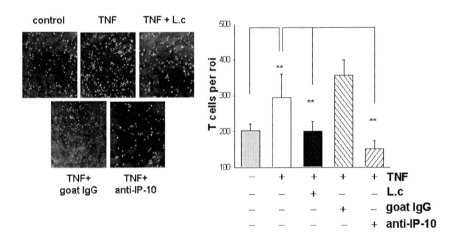

Figure 10: *L. casei* (VSL#3)-mediated selective reduction of IP-10 in the supernatant of TNF-activated IEC is sufficient to block IEC-induced T cell recruitment
The left figure shows representative pictures of transmigrated activated T lymphoblasts and the figure on the right shows the mean number of transmigrated activated T cells per region of interest (roi) after a T cell transmigration assay using the indicated IEC-conditioned media. Neutralisation of IP-10 in the supernatant of TNF-activated IEC using a goat-anti-IP-10 antibody resulted in significant reduction of the number of transmigrated T cells, whereas a control-goat antibody did not have similar effects.

4.1.3 Surface proteins of *L. casei* (VSL#3) induce the inhibition of IP-10 in IEC via TLR2-independent mechanisms

To unravel the question, which molecular structures of *L. casei* (VSL#3) are responsible for the observed inhibition of IP-10 expression in IEC, the probiotic bacteria was subjected to heat, formaldehyde-fixation and enzymatic treatments. The effect of these treatments on *L. casei* (VSL#3)-mediated inhibition of IP-10 expression was subsequently analysed in cell culture stimulation experiments. The results show that active interaction between viable *L. casei* (VSL#3) and IEC is not necessary for the inhibition of IP-10, as formaldehyde-fixed bacteria were as effective as live bacteria. In contrast, heat treatment or lysis of *L. casei* (VSL#3) by lysozyme completely abrogated the inhibitory effect. These results show that heat-labile surface structures are the active bacterial component. In an attempt to clarify the nature of this surface component, enzymatic treatment studies with fixed *L. casei* (VSL#3) were performed. It was found that phospholipase A treatment did not affect the inhibitory effect of *L. casei* (VSL#3) on IP-10. In contrast, Proteinase K as well as trypsin treatment of *L. casei* (VSL#3) abolished its ability to inhibit IP-10 expression in IEC (Figure 11), revealing that the active component is a surface protein of *L. casei* (VSL#3). As a consequence of these findings, formaldehyde fixed bacteria were used in all *in vitro* studies, allowing better standardisation of the experiments.

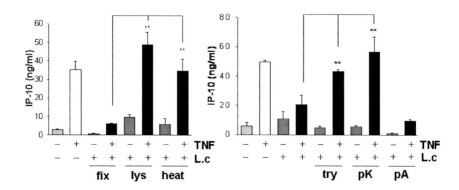

Figure 11: Protease-sensitive surface components of *L. casei* (VSL#3) are the active bacterial structures with regard to the inhibition of IP-10 in IEC.
ELISA analysis of cell culture supernatants after stimulation experiments with differentially treated *L. casei* (VSL#3) showed that live bacteria are not necessary for the observed inhibition of IP-10 expression in IEC. In contrast, proteolytic treatment abrogates the anti-inflammatory effect of *L. casei* (VSL#3). Abbreviations: fix; fixed, lys; lysed by lysozyme, heat; heat-killed, try; trypsin, pK: proteinase K, pA; phospholipase A

In consistence with the observation that the active bacterial component is a specific surface protein of *L. casei* (VSL#3), and not a ubiquitous cell wall component of gram-positive bacteria, the inhibition of IP-10 in IEC was found to be independent of the pattern recognition receptor TLR2. Probiotic stimulation was found to inhibit IP-10 secretion to the same extent in TLR2-negative as well as TLR2-positive HEK cells (Figure 12). Of note, the finding that *L. casei* (VSL#3) inhibits IP-10 secretion not only in Mode K cells but also in HEK cells shows that the inhibitory mechanism that is induced by *L. casei* (VSL#3) is not cell line specific or restricted to IEC.

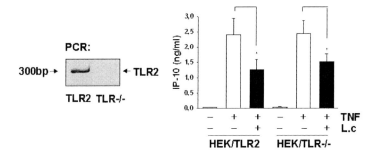

Figure 12: The selective inhibition of IP-10 secretion in IEC is independent of TLR2-dependent signalling pathways.
The left figure shows the result of a TLR2-specific PCR that was performed after reverse transcription of mRNA isolated from the indicated HEK cells, confirming lack of TLR2 expression in TLR2-/- HEK cells. ELISA analysis (right figure) of stimulation experiments in TLR2-positive or TLR2-deficient HEK cells revealed that the inhibition of IP-10 by *L. casei* (VSL#3) is independent of TLR2-dependent signalling pathways.

4.1.4 *L. casei* (VSL#3) does not inhibit TNF-induced NFκB signalling

TNF-induced IP-10 expression is mediated via the activation of the transcription factor NFκB. As VSL#3 had previously been shown to inhibit NFκB activation by the inhibition of proteasome activity (Petrof et al., 2004), it was speculated that *L. casei* (VSL#3) might mediate the observed inhibition of IP-10 via analogous effects. Surprisingly, stimulation experiments with Mode K cells revealed that *L. casei* (VSL#3) did not inhibit proteasome activity. Lactacystin, a known proteasome inhibitor was used as a positive control. Consistently, neither TNF-induced degradation of IκB nor TNF-induced phosphorylation of RelA/p65 was inhibited by *L. casei* (VSL#3) (Figure 13).

A   B

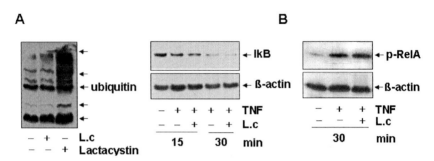

Figure 13: TNF-induced NFκB signalling is unaffected by stimulation with *L. casei* (VSL#3).
Western Blot analysis of untreated cells, *L. casei* (VSL#3)-stimulated cells and lactacystin-treated cells
(positive control) showed that *L. casei* (VSL#3) did not inhibit proteasomal degradation in Mode K cells
(A). TNF-induced IκB degradation (A) and TNF-induced RelA phosphorylation (B) were also not
inhibited by *L. casei* (VSL#3).

4.1.5 *L. casei* (VSL#3) inhibits TNF-induced IP-10 expression at a post-transcriptional level

In order to investigate whether *L. casei* (VSL#3) interferes with the recruitment of RelA to the
IP-10 promoter or with IP-10 promoter activity in general, chromatin immunoprecipitation
(ChIP) assays as well as luciferase reporter assays were performed. It was found that
L. casei (VSL#3) did not block TNF-induced recruitment of RelA to the IP-10 promoter and
that TNF-induced IP-10 promoter activity was also not reduced by the probiotic bacteria
(Figure 14).

A   B

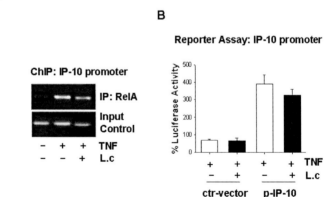

Figure 14. TNF-induced IP-10 promoter activity is unaffected by *L. casei* (VSL#3).
ChIP analysis demonstrated that TNF-induced recruitment of RelA to the IP-10 promoter is not
inhibited by *L. casei* (VSL#3) (A). TNF-induced luciferase reporter activity measured within Mode K
cell lysates was also not reduced by *L. casei* (VSL#3) (B).

Subsequent IP-10 mRNA analysis of TNF and/or *L. casei* (VSL#3)-stimulated Mode K cells six hours after the stimulation showed that *L. casei* (VSL#3) did not inhibit TNF-induced IP-10 mRNA transcription. In contrast to the lack of effects of probiotic stimulation on NFκB signal transduction and IP-10 gene transcription, intracellular IP-10 protein was almost completely lost in IEC after 24 h of costimulation (Figure 15). This finding suggests that *L. casei* (VSL#3) induces inhibitory mechanisms that mediate the loss of IP-10 on a post-transcriptional level.

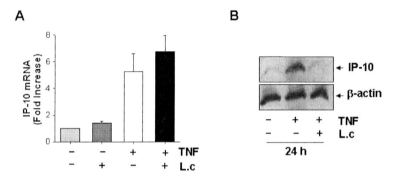

Figure 15: *L. casei* (VSL#3) mediates loss of IP-10 protein via post-transcriptional mechanisms in IEC.
Real-time PCR analysis showed that IP-10 mRNA levels are not reduced after probiotic stimulation (6h) (A) whereas Western Blot analysis showed that IP-10 protein is strongly reduced within IEC (24 h) (B).

The finding that *L. casei* (VSL#3) induces post-transcriptional inhibitory mechanisms raised the question of whether the observed loss of IP-10 might be completely independent of TNF-induced signal transduction pathways. To prove this hypothesis, the effects of *L. casei* (VSL#3) on IFNγ-induced IP-10 expression in IEC were investigated. In addition, IP-10 overexpression experiments were performed using HEK cells instead of Mode K cells, as the former ones show a much higher transfection efficacy. HEK cells were transfected with plasmids encoding for an IP-10-DsRed fusion protein (IP-10-DsRed), or control plasmids, encoding for the reporter protein DsRed, under the control of a constitutively active murine leukemia virus (MLV)-promoter. *L. casei* (VSL#3) was found to inhibit IFNγ-induced IP-10 expression as well as the amount of constitutively-produced IP-10 in IEC (Figure 16). These results clearly show that the post-transcriptional inhibition of IP-10 by *L. casei* is stimuli independent.

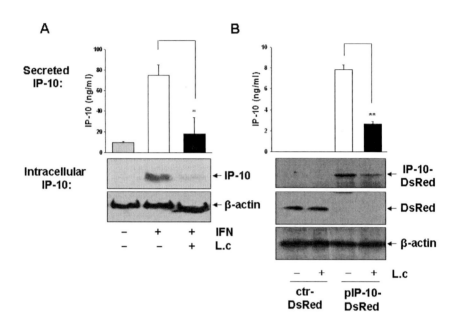

Figure 16. The inhibition of IP-10 in IEC by *L. casei* (VSL#3) is stimuli independent.
ELISA analysis showed that *L. casei* (VSL#3) inhibits IFNγ-induced IP-10 secretion into the cell culture supernatant (upper panel) and Western Blot analysis revealed analogous loss of IP-10 within IEC (lower panel) (A). In addition, constitutively expressed IP-10 protein is lost within IEC, as well as in the cell culture supernatant, after stimulation with *L. casei* (VSL#3), whereas overexpression of a control protein (DsRed) is unaffected by probiotic treatment (B).

4.1.6 *L. casei* (VSL#3) induces an IP-10 specific secretional blockade resulting in degradation of the chemokine in IEC

In order to reveal whether *L. casei* (VSL#3) had an effect on protein translation in IEC, the activation status of two important proteins of the translational machinery, EIF4G and EIF4E, was analysed after the stimulation with TNF or TNF and *L. casei* (VSL#3). *L. casei* (VSL#3) did not inhibit TNF-induced activation of EIF4G and EIF4E. In contrast, the probiotic bacteria was found to increase TNF-induced activation of EIF4G and EIF4E (Figure 17), which is consistent with increased expression of IL-6 (Figure 7) and MIP-2 (data not shown) in IEC that were costimulated with TNF and *L. casei* (VSL#3). Kinetic analysis of IP-10 expression in TNF-activated Mode K cells revealed that IP-10-specific protein translation was also not inhibited by *L. casei* (VSL#3). IP-10 protein was found to be initially produced but it was lost intracellularly at later time points (6h, 24h) in the presence of *L. casei* (VSL#3) (Figure 17).

However, the observed loss of IP-10 protein in IEC was not due to secretion of IP-10 into the cell culture supernatant, suggesting that *L. casei* (VSL#3) induces post-translational degradation of IP-10 in IEC.

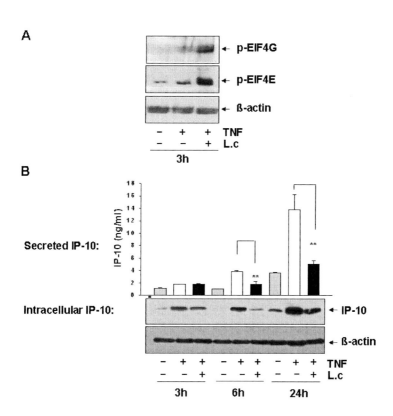

Figure 17. *L. casei* (VSL#3) does not inhibit initial IP-10 translation in IEC

Western Blot analysis showed that *L. casei* (VSL#3) does not inhibit the activation of the translational machinery in IEC (A). In addition, initial production of IP-10 protein is not inhibited by *L. casei* (VSL#3) whereas intracellular IP-10 protein is lost at later time points without being secreted into the cell culture supernatant (B).

To investigate the fate of initially produced IP-10 protein in IEC in the presence of *L. casei* (VSL#3), pulse chase experiments were performed using S^{35}-labelling of proteins. Subsequent immunoprecipitation of IP-10 revealed that initially produced S^{35}-labelled IP-10 protein was absent within cells that were not stimulated with *L. casei* (VSL#3) during the chase period, indicating normal secretion of the chemokine in these cells. In contrast, S^{35}-

labelled IP-10 protein was still present in IEC that were stimulated with *L. casei* (VSL#3) during the chase period (Figure 18), suggesting that the probiotic bacteria mediates inhibition of IP-10 secretion. Interestingly, the inhibition of the secretory machinery via Brefeldin A resulted in loss of IP-10 analogous to the one induced by *L. casei* (VSL#3) (Figure 18). This finding led to the hypothesis, that initial IP-10 accumulation due to a secretory blockade might induce an intrinsic degradation program in IEC, in order to prevent harmful chemokine accumulation.

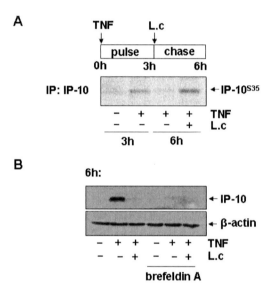

Figure 18. *L. casei* (VSL#3) mediates an IP-10-specific secretional blockade and subsequent activation of degradation mechanisms resulting in loss of IP-10 protein.
Pulse-chase experiments showed that TNF-induced S^{35}-labelled IP-10 protein, that is produced during the 3h pulse period, is captured within *L. casei* (VSL#3)-stimulated IEC throughout the following 3h chase period. In contrast, S^{35}-labelled IP-10 protein is lost in unstimulated IEC due to normal protein secretion (A). Western Blot analysis showed that the inhibition of protein secretion via brefeldin A also results in loss of IP-10 protein (B).

The finding that IP-10 protein shows a distinct ubiquitination pattern after six hours of costimulation with TNF and *L. casei* (VSL#3) (Figure 19) suggested that ubiquitine-dependent degradation pathways might be involved in the observed loss of IP-10 in IEC. Surprisingly, the blockade of proteasomal activity by lactacystin, or the blockade of lysosomal degradation by NH$_4$Cl, did not rescue IP-10 protein (Figure 19).

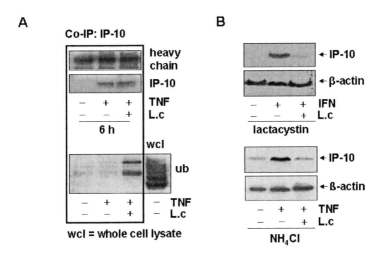

Figure 19. Degradation of IP-10 protein in IEC is dependent on alternative degradation pathways.
Immunoprecipitation of IP-10 and subsequent Western Blot analysis showed that IP-10 is ubiquitinated after 6 h of stimulation with *L. casei* (VSL#3) (A). Ubiquitination does not result in IP-10 degradation via proteasomal or lysosomal pathways, as the inhibition of these pathways by lactacystin or ammoniumchloride (NH₄Cl) does not prevent IP-10 degradation (B).

The blockade of the third major cellular degradation machinery, autophagy, by 3-methyladenine (3-MA) did also not inhibit IP-10 degradation. Mechanistically of most interest, the inhibition of autophagic vesicle formation by 3-MA resulted in similar effects as the stimulation with *L. casei* (VSL#3) with regard to cytokine and chemokine secretion. The blockade of autophagy-related vesicular transport was found to result in loss of intracellular and secreted IP-10, whereas the production and secretion of IL-6 was found to be even increased. In addition, the loss of IP-10 was found to be mediated via post-transcriptional mechanisms (Figure 20). This result indicates that IP-10 secretion is dependent on the formation of autophagy-related vesicles, suggesting that *L. casei* (VSL#3) might have inhibitory effects on this specific IP-10 secretory pathway.

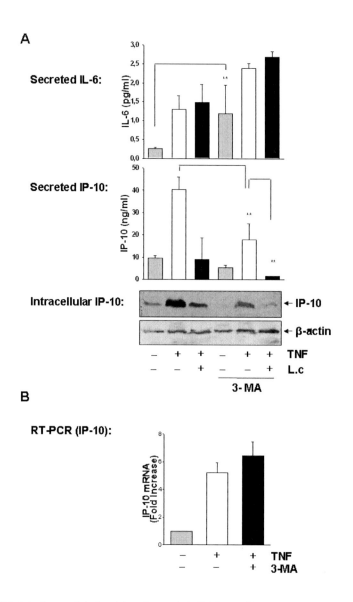

Figure 20. Autophagy-related vesicles play a role in IP-10 secretion.

The inhibition of vesicle formation by 3-MA results in exactly the same selective inhibition of IP-10 secretion as it was observed after the stimulation with *L. casei* (VSL#3). In addition, 3-MA mediates loss of intracellular IP-10 protein after 24 h of stimulation (A) without affecting TNF-induced IP-10 gene transcription (6 h) (B).

4.2 Lactocepin, a surface-associated and secreted bacterial serin protease is responsible for the observed loss of IP-10

4.2.1 Secreted compounds of *L. casei* (VSL#3) induce loss of IP-10 protein in IEC analogous to bacterial surface proteins

In order to analyse the effects of secreted components of *L. casei* (VSL#3) on TNF-activated IEC, we generated bacterial conditioned media (CM) of *L. casei* (VSL#3) (CM L.c). Interestingly, we found that *L. casei* (VSL#3) does secrete active components that induce selective inhibition of IP-10 secretion in IEC (Figure 21), analogous to bacterial cell surface components (Figure 11). TNF-induced IL-6 secretion (Figure 21) and MIP-2 secretion (data not shown) are not reduced by the stimulation of IEC with CM *L. casei* (VSL#3).

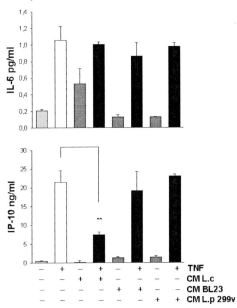

Figure 21. CM of *L. casei* (VSL#3) selectively inhibits IP-10 secretion whereas CM of *L. plantarum* 299v and *L. casei* BL23 do not show analogous effects.

ELISA analysis revealed that CM of *L. casei* (VSL#3) reduces the concentration of IP-10 in the supernatant of TNF-activated IEC whereas it induces the secretion of IL-6.

The observed inhibition of IP-10 via bacterial CM was found to be specific for *L. casei* (VSL#3) as CM generated by *L. plantarum* 299v (CM L.p 299v) or *L. casei* BL23 (CM BL23) did not exert similar effects (Figure 21). Of note, CM generated by *E. coli* Nissle strongly induced apoptosis in IEC whereas CM generated by the gram-positive bacterial strains did not affect viability of IEC (data not shown). To clarify whether the observed inhibition of IP-10 by secreted compounds of *L. casei* (VSL#3) was mediated via the same molecular mechanisms than the one observed with whole cell *L. casei* (VSL#3), several central experiments were repeated with CM *L. casei* (VSL#3) using CM *L. casei* BL23 as a negative control. Analogous to cell surface components of *L. casei* (VSL#3), CM *L. casei* (VSL#3) induced almost complete loss of IP-10 protein in IEC whereas TNF-induced IP-10 mRNA expression and IκB degradation were found to be unaffected (Figure 22).

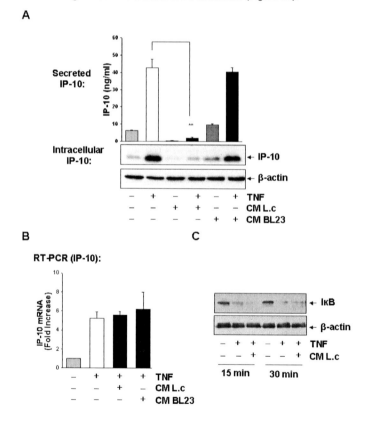

Figure 22. CM *L. casei* (VSL#3) induces loss of intracellular IP-10 protein via a post-transcriptional mechanism analogous to whole cell *L. casei* (VSL#3).
Western Blot analysis showed that intracellular IP-10 protein is completely lost after stimulation with CM *L. casei* (VSL#3) without being secreted into the cell culture supernatant. CM *L. casei* BL23 did not have analogous effects (A). RT-PCR analysis revealed that TNF-induced IP-10 gene expression (B) and TNF-induced IκB degradation are unchanged after stimulation with CM *L.casei* (VSL#3) (C).

Consistent with previous experiments, CM *L. casei* (VSL#3)-mediated loss of IP-10 protein in IEC was found to be independent of the initial stimuli that induced IP-10 expression. IFNγ-induced IP-10 protein accumulation in IEC was strongly inhibited by CM *L. casei* (VSL#3) (Figure 23).

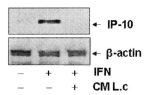

$\leftarrow$ **IP-10**

$\leftarrow$ **β-actin**

| − | + | + | **IFN** |
| − | − | + | **CM L.c** |

Figure 23. CM *L. casei* (VSL#3)-induced loss of IP-10 protein is stimuli independent.
CM *L. casei* (VSL#3) strongly inhibits IFNγ-induced increase of intracellular IP-10 protein.

In summary, these results indicate that the active secreted components of *L. casei* (VSL#3) and the active *L. casei* (VSL#3) surface proteins mediate analogous post-translational mechanisms, resulting in loss of IP-10 protein in IEC. This finding suggests, that the surface-associated and the secreted active components of *L. casei* (VSL#3) might be identical bacterial structures.

4.2.2 The active bacterial component secreted by *L. casei* (VSL#3) is a bacterial protease

Bacterial CM is a very complex suspension of abundant bacterial components like unmethylated DNA, metabolites, proteins and polysaccharides. In order to analyse, which specific bacterial component is responsible for the observed inhibition of IP-10 expression in IEC, CM was subjected to several treatments. Heat-treatment of CM *L. casei* (VSL#3) was found to result in inactivation of the active component (Figure 24) whereas Benzonase A treatment (data not shown) did not have any effect, excluding bacterial DNA as the active component. Activity was found to be precipitable by 50% ammoniumsulfate (Figure 24), suggesting that the inhibition of IP-10 secretion is mediated via proteinaceous bacterial components. This result was consistent with the fact that the active surface-expressed components of *L. casei* (VSL#3) were also found to be proteinaceous in nature (Figure 11).

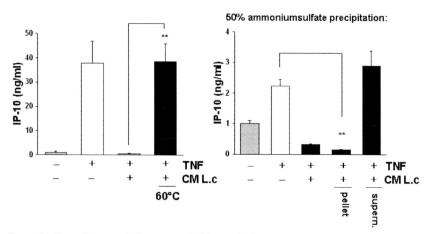

Figure 24. The active secreted component of *L. casei* (VSL#3) is a bacterial protein.
ELISA analysis with differently treated CM *L. casei* (VSL#3) revealed that heating of CM *L. casei* (VSL#3) abrogates its inhibitory activity on IP-10 secretion, whereas 50% ammoniumsulfate precipitation results in complete precipitation of the active component. Activity is regained by resolubilisation of the precipitate whereas the precipitate supernatant does not exert any detectable activity on IP-10 secretion.

Importantly, size exclusion experiments showed that the active protein is bigger than 50 kDa. The active component is concentrated in the retentate using a 50 kDa filter device, whereas the flow-through was found to be completely unactive (Figure 25).

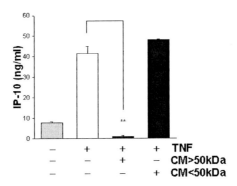

Figure 25. The active secreted bacterial protein is bigger than 50 kDa.
Concentration of CM *L. casei* (VSL#3) using a 50 kDa filter device results in loss of activity in the flow-through whereas the retentate (1x) remains active.

Surprisingly, PMSF, an irreversible serine/cysteine protease inhibitor, abrogated the inhibitory effect of CM *L. casei* (VSL#3) on IP-10 expression in IEC (Figure 26). This result raised the question of whether the PMSF-mediated effect was due to the inhibition of an active bacterial protease or due to the inhibition of a cellular protease involved in the degradation of IP-10 in IEC. Stimulation experiments with untreated CM *L. casei* (VSL#3) in PMSF-preincubated Mode K cells revealed that PMSF does not inhibit the degradation of IP-10 in IEC, but that it inhibits the activity of a bacterial protease in the CM of *L. casei* (VSL#3) (Figure 26).

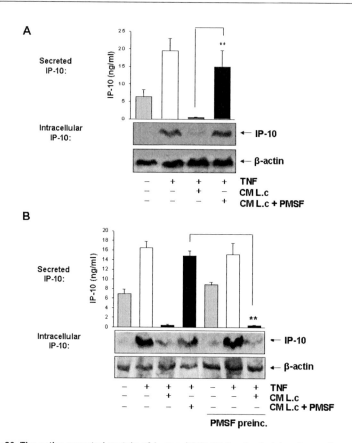

Figure 26. The active secreted protein of *L. casei* (VSL#3) is a bacterial serine protease.
The application of the irreversible serine protease inhibitor PMSF completely abrogates the inhibitory effect of CM *L. casei* (VSL#3) on IP-10 protein secretion (upper panel) and rescues intracellular IP-10 protein (lower panel) (A). The observed inhibition of IP-10 degradation is due to the inhibition of a bacterial protease and not due to the inhibition of a cellular protease, as PMSF-preincubation (1 h) of Mode K cells does not inhibit subsequent induction of IP-10-degradation via CM *L. casei* (VSL#3) (B).

Subsequent serine/cystein protease activity assays showed that CM *L. casei* (VSL#3) indeed contains one or more PMSF-sensitive serine/cysteine proteases in contrast to CM *L. casei* BL23. However, the proteolytic activity of CM *L. casei* (VSL#3) in the FTC-casein assay was found to be rather weak and was only observed when CM *L. casei* (VSL#3) was used in high concentrations (Figure 27). These results suggest that FTC-casein is not a preferred substrate for the active bacterial protease that mediates the loss of IP-10 in IEC.

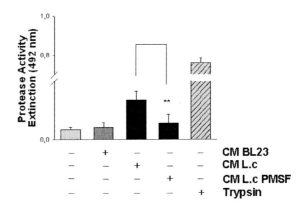

−	+	−	−	−	**CM BL23**
−	−	+	−	−	**CM L.c**
−	−	−	+	−	**CM L.c PMSF**
−	−	−	−	+	**Trypsin**

Figure 27. L. casei (VSL#3), but not L. casei BL23 secretes a bacterial serine protease which is able to degrade FTC-casein.
CM L. casei (VSL#3) (10x) is able to degrade FTC-casein whereas CM L. casei BL23 (10x) does not show proteolytic activity.

4.2.3 The active bacterial protease does not reduce TEER

Bacterial proteases were recently shown to be able to reduce intestinal barrier function (Steck et. al, 2009, Gastroenterology, 136-5, supplement 1, A21-A22). In order to investigate whether the secreted protease of L. casei (VSL#3) exerts similar detrimental effects on IEC barrier function, T84 cells were stimulated with CM L. casei (VSL#3) using CM L. casei BL23 as negative control. The active protease secreted by L. casei (VSL#3) did not reduce TEER of T84 cells. In addition, CM L. casei (VSL#3) did not induce further decrease of TNF/IFNy-reduced TEER of T84 cells (Figure 28). This result shows that the active bacterial protease secreted by L. casei (VSL#3) does not negatively influence intestinal epithelial barrier function.

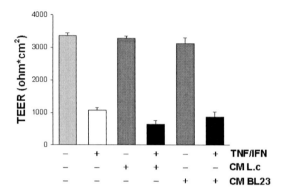

Figure 28. The anti-inflammatory bacterial protease secreted by *L. casei* (VSL#3) does not negatively affect TEER.

The stimulation of T84 cells with CM *L. casei* (VSL#3) and CM *L. casei* BL23 revealed that the active bacterial protease in the CM *L. casei* (VSL#3) does not affect normal or TNF/IFNγ-reduced TEER.

4.2.4 Lactocepin is the active bacterial protease of *L. casei* (VSL#3)

In order to characterize the active bacterial protease that is secreted by *L. casei* (VSL#3), chromatographic fractionation of CM *L. casei* (VSL#3) was performed in collaboration with Dr. Alpert at the DifE in Potsdam. CM *L. casei* (VSL#3) was fractionated according to molecular mass (superdex chromatography), hydrophobicity (phenyl-sepharose) or charge (MonoQ-ion exchange chromatography). The resulting chromatographic fractions were subsequently screened in cell culture stimulation experiments with regard to their potential to reduce IP-10 secretion. Surprisingly, each of the three different chromatographic methods led to the elution of several active fractions, which were separated by non-active fractions (Figure 29). This uncommon elution profile of the active protein points towards an oligomeric protease with different active subunits, self-processing of the active protease during elution or the existence of several independently active proteases in the CM of *L. casei* (VSL#3).

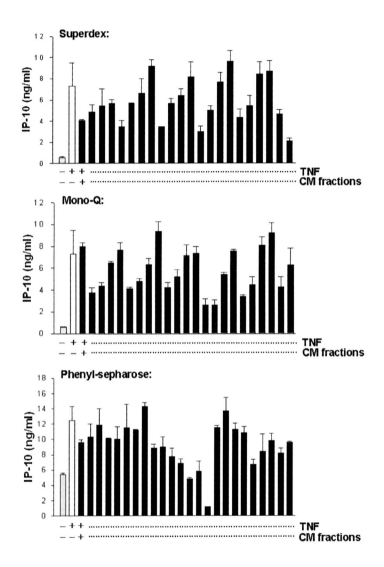

Figure 29. Chromatographic fractionation of CM _L. casei_ (VSL#3) results in several separated active fractions.

Chromatographic separation of the constituents of CM _L. casei_ (VSL#3) according to molecular mass (Superdex), charge (Mono-Q) or hydrophobicity (Phenyl-Sepharose) was performed. Subsequent cell culture stimulation experiments revealed probiotic activity in several chromatographic fractions which were often separated by non-active fractions.

Several chromatographic fractions that were identified to be highly active or non-active were subsequently subjected to LC-MS-MS analysis. All active fractions were found to contain a bacterial protease, lactocepin, that was absent from non-active fractions (Table 4). Lactocepin is a membrane-bound and secreted serine protease with a molecular weight of 200 kDa (precursor), thereby fitting all the previously detected characteristics of the active component of *L. casei* (VSL#3). This result suggests that lactocepin is the active probiotic structure that mediates the observed loss of IP-10 in IEC.

Table 4: Lactocepin was identified as the active protease secreted by *L. casei* (VSL#3).

gene name	protein name	MW (kDa)	chromatography fractions					
			n.a	a	a	n.a	a	n.a
l	D-2 hydroxyisocaproate dehydrogenase	36	+	+				
groL	60kDa chaperonin	57	+	+	+		+	
valS	Valyl-tRNA synthethase	100	+	+				
gap	Glyceraldehyde-3-phosphate dehydrogenase	36	+	+	+	+		+
queA	S-adenosylmethionine:tRNA ribosyltransferase	38	+					
gnd	6-phosphogluconate dehydrogenase	52	+	+				
groS	10 kDa Chaperonin	10		+	+			
prtP	**PII-type proteinase precursor (Lactocepin)**	**200**		+	+		+	
metG	MethionyltRNA synthetase	76				+		
dnaK	hsp70	66				+		
ldh	L.lactatdehydrogenase	35					+	
galE	UDP-glucose 4-epimerase	36					+	
prtP	PI-type proteinase precursor	199					+	
purF	Amidophosphoribosyltransferase	112					+	
guaB	Inosine-5'-monophosphate dehydrogenase	52					+	
pgk	Phosphoglycerate kinase	42						+
tuf	Elongation factor Tu	43						+

n.a (not active)
a (active)

Different active as well as non-active fractions derived from chromatographic separations were subjected to LC-MS-MS analysis. The table shows all proteins that were identified in the respective fractions. All active fractions were found to contain a cell wall associated and secreted serine protease, lactocepin, which was absent from the non-active fractions.

4.3 VSL#3 and *L. casei* (VSL#3) are not effective in the prevention or treatment of ileitis in TNF$^{\Delta ARE/+}$ mice

4.3.1 Post-weaning feeding of VSL#3 or *L. casei* (VSL#3) does not reduce ileitis in TNF$^{\Delta ARE/+}$ mice

In order to analyse whether the inhibition of IP-10 secretion in IEC by VSL#3 and *L. casei* (VSL#3) is a physiologically relevant anti-inflammatory mechanism with regard to ileal inflammation, we performed feeding experiments in TNF$^{\Delta ARE/+}$ mice, an experimental model of chronic ileitis. Feeding of TNF$^{\Delta ARE/+}$ mice with VSL#3 or *L. casei* (VSL#3) was started post-weaning. At the age of 18 weeks, the mice were sacrificed and samples were taken for subsequent analysis of probiotic effects. Histopathological analysis of distal ileal sections showed that VSL#3 did not have protective effects on ileitis in TNF$^{\Delta ARE/+}$ mice. Analogously, *L. casei* (VSL#3) was not effective in the reduction of ileitis in TNF$^{\Delta ARE/+}$ mice (Figure 30).

A Ileum

wt + placebo   TNF$^{\Delta ARE/+}$ + VSL#3

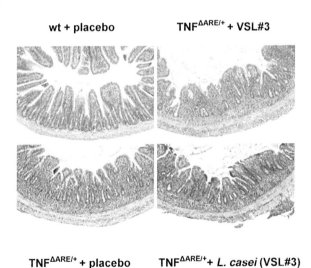

TNF$^{\Delta ARE/+}$ + placebo   TNF$^{\Delta ARE/+}$+ *L. casei* (VSL#3)

B Ileum

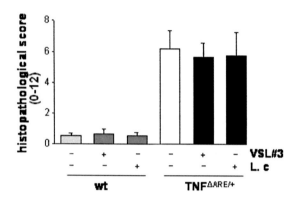

Figure 30. VSL#3 and _L. casei_ (VSL#3) do not reduce ileitis in TNF$^{\Delta ARE/+}$ mice.
The figure shows representative H&E stained ileal tissue sections (A) and the mean histopathological scores (B) of differentially treated wildtype and TNF$^{\Delta ARE/+}$ mice, demonstrating that none of the two probiotic treatments had a protective effect on ileal inflammation.

As the probiotic treatments were uneffective, it was important to prove that the applied feeding protocol did result in an increase of probiotic bacteria in the intestine of TNF$^{\Delta ARE/+}$ mice. PCR analysis was performed with DNA isolated from gut content using species-specific primers for _S. thermophilus_, the most prominent bacterial strain of VSL#3, or _L. casei_ (VSL#3) (Figure 31). As expected, this analysis confirmed increased numbers of _S. thermophilus_ in the intestine of VSL#3-fed mice as well as increased numbers of _L. casei_ (VSL#3) in the intestine of _L. casei_ (VSL#3)-fed mice. Interestingly, _L. casei_ (VSL#3) was also found to be attached to the ileal mucosa in _L. casei_ (VSL#3)-fed mice.

A

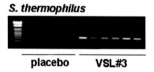

B

L. casei

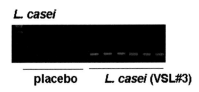

placebo   L. casei (VSL#3)

L. casei (cfu/10 mg tissue):

	placebo	L. casei (log10)
wildtype	n.d.	4.5 ± 1.3
TNFdeltaARE	n.d.	5.3 ± 1.1

Figure 31. VSL#3 and *L. casei* (VSL#3) are present in the intestine of TNF$^{\Delta ARE/+}$ mice.

The numbers of *S. thermophilus* (A) and *L. casei* (VSL#3) (B) are increased in the gut content of VSL#3-and *L. casei* (VSL#3)-fed mice compared to placebo-fed mice. The table in Figure B shows that *L. casei* (VSL#3) is even attached to the ileal mucosa in detectable amounts in *L. casei* (VSL#3)-fed mice whereas this is not the case in placebo fed mice.

In consistence with the lack of effects on ileal inflammation in TNF$^{\Delta ARE/+}$ mice, the level of IP-10 in isolated ileal/jejunal IEC was not reduced by the different probiotic treatments (Figure 32). Furthermore, subsequent proteome analysis revealed that VSL#3-feeding did not change protein expression profiles in primary ileal/jejunal IEC isolated from TNF$^{\Delta ARE/+}$ mice (data not shown).

Ileal/jejunal IEC:

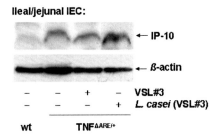

Figure 32: Ileal IEC isolated from inflamed TNF$^{\Delta ARE/+}$ mice show increased levels of IP-10 protein which were not reduced by probiotic treatments.

Western Blot analysis of isolated primary ileal/jejunal epithelial cells revealed that IP-10 protein levels are strongly upregulated under inflammatory conditions and that the uptake of VSL#3 or *L. casei* (VSL#3) did not counteract this effect.

4.3.2 Combined pre-weaning and post-birth VSL#3-feeding does not prevent ileitis in TNF$^{\Delta ARE/+}$ mice

In order to analyse whether earlier onset of probiotic therapy would be effective in the reduction or prevention of ileitis in TNF$^{\Delta ARE/+}$ mice, a pre-weaning feeding study was performed. Pregnant wildtype mice were fed VSL#3 and in addition, the offspring of these mice (wildtype and TNF$^{\Delta ARE/+}$ mice) were fed VSL#3 from day one after birth. The mice were sacrificed at the age of 18 weeks. Histopathological analysis of the distal ileum revealed that the probiotic mixture did not exert protective effects on the ileitis in TNF$^{\Delta ARE/+}$ mice in this experimental setup (Figure 33). As plasma levels of serum amyloid A (SAA), a general marker of inflammation, were paradoxically found to be undetectable even in highly inflamed TNF$^{\Delta ARE/+}$ mice, plasma levels of TNF were determined in order to assess the systemic impact of VSL#3. In addition to the lack of protective effects on ileal inflammation, plasma levels of TNF were found to be unaffected by probiotic treatment in TNF$^{\Delta ARE/+}$ mice (data not shown).

A   Ileum

wt + placebo   TNF$^{\Delta ARE/+}$+ placebo

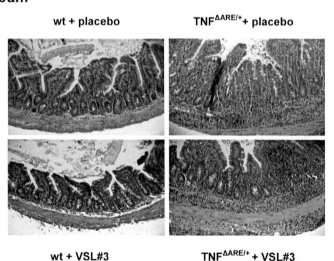

wt + VSL#3   TNF$^{\Delta ARE/+}$ + VSL#3

B   Ileum

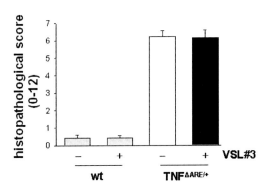

Figure 33: Pre-weaning feeding of VSL#3 does not prevent ileal inflammation in TNF$^{\Delta ARE/+}$ mice.
The figure shows representative pictures of H&E stained ileal tissue sections (A) as well as the mean histopathological score (B) of placebo or probiotic-treated wildtype and TNF$^{\Delta ARE/+}$ mice, demonstrating that pre-weaning feeding of VSL#3 has no protective effect on ileal inflammation.

4.4   Protective effects of VSL#3 on colitis in IL10-/- mice are segment-specific and correlate with IP-10 expression in IEC

4.4.1 VSL#3 is protective on cecal inflammation in IL10-/- mice whereas colonic inflammation is unaffected by probiotic treatment

Previous studies had already shown significant reduction of cecal and colonic inflammation in IL-10-/- mice by VSL#3 (Madsen et al., 2001). In order to analyse the protective effects of VSL#3 on IEC in this experimental colitis model, IL-10-/- and 129SvEv wildtype mice were fed VSL#3, starting post-weaning. At 24 weeks of age the mice were sacrificed and samples were taken for subsequent analysis of probiotic effects. Surprisingly, histopathological analysis of the cecum and the distal colon showed that VSL#3 had segment-specific protective effects as colonic inflammation was not reduced whereas cecal inflammation was significantly abrogated in probiotic fed IL10-/- mice (Figure 34).

A Colon

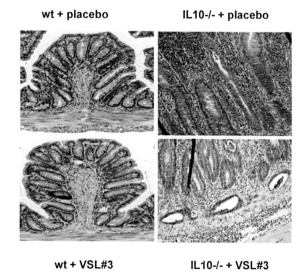

wt + placebo   IL10-/- + placebo

wt + VSL#3   IL10-/- + VSL#3

B Colon

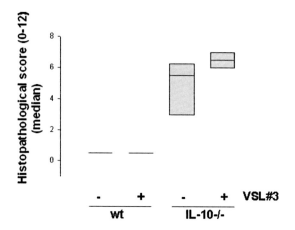

C Cecum

wt + placebo   IL10-/- + placebo

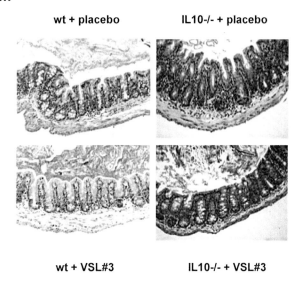

wt + VSL#3   IL10-/- + VSL#3

D Cecum

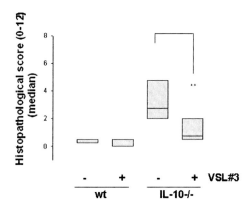

Figure 34. VSL#3 has local protective effects on cecal inflammation in IL10-/- mice.

Representative pictures of H&E stained colonic tissue sections (A) and the median histopathological scores (B) of the colon of placebo or VSL#3-fed wildtype and IL10-/- mice demonstrate that VSL#3 had no protective effect on colonic inflammation. In contrast, representative pictures of H&E stained cecal tissue sections (C) and the median histopathological scores of the cecum (D) of placebo or VSL#3-fed wildtype or IL10-/- mice demonstrate that VSL#3 reduces cecal inflammation in IL10-/- mice.

Systemically, enhanced serum amyloid A (SAA) plasma levels were found to be unaffected by probiotic feeding (Figure 35), underlining the histopathological finding that the anti-inflammatory effect of VSL#3 on IL10-/- mice was due to a local probiotic mechanism restricted to the cecum.

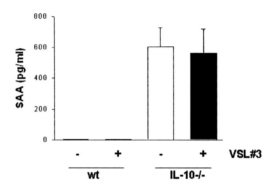

Figure 35: The systemic inflammation marker SAA is not reduced by VSL#3 feeding

ELISA analysis showed that SAA is strongly increased in the plasma of IL10-/- mice and that probiotic feeding did not reduce the high plasma level of SAA.

4.4.2 VSL#3-induced attenuation of cecal inflammation correlated with reduced expression of IP-10 in cecal epithelial cells

Protein analysis of pooled cecal/colonic IEC revealed reduced levels of IP-10 in IEC isolated from VSL#3-fed IL10-/- mice compared to the ones isolated from placebo-fed IL10-/- mice. Immunohistochemical analysis of cecal tissue sections confirmed that IP-10 expression was strongly reduced in cecal epithelial cells of VSL#3-fed IL10-/- mice (Figure 36) whereas there was no difference in IP-10 expression in colonic IEC (data not shown). In consistence with the reported post-translational inhibition of IP-10 expression in IEC *in vitro*, the reduction of IP-10 protein in pooled primary cecal/colonic IEC was found to be mediated via post-translational mechanisms, as the production of IP-10 mRNA in these cells was not reduced by VSL#3 (Figure 36). These results suggest that the local induction of IP-10 degradation in cecal IEC by *L. casei* (VSL#3) might play an important role in the observed VSL#3-mediated reduction of cecal inflammation in IL10-/- mice.

A

Colonic/cecal epithelial cells:

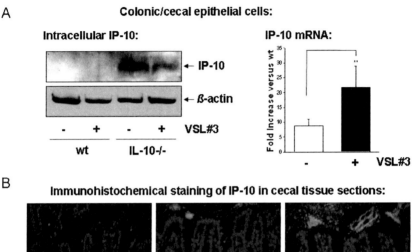

Intracellular IP-10:

← IP-10

← ß-actin

- + - + VSL#3

wt   IL-10-/-

IP-10 mRNA:

B

Immunohistochemical staining of IP-10 in cecal tissue sections:

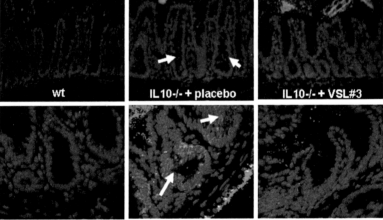

wt   IL10-/- + placebo   IL10-/- + VSL#3

Figure 36: Reduced cecal inflammation is paralleled by reduced IP-10 protein expression in cecal epithelial cells.
Western Blot analysis of isolated cecal/colonic epithelial cells revealed a strong increase of IP-10 in IEC isolated from placebo-fed IL10-/- mice. VSL#3-feeding was found to decrease the level of IP-10 protein in IEC whereas IP-10 mRNA expression within these cells was not reduced (A). Immunohistochemical staining (B) revealed that IP-10 protein (green) is strongly reduced in cecal epithelial cells of VSL#3-fed mice, correlating with reduced cecal inflammation.

DISCUSSION

The aim of the present thesis was to reveal bacterial structure-related molecular mechanisms underlying protective effects of VSL#3 in the context of IBD. The findings revealed post-translational inhibition of IP-10 secretion in IEC as a new probiotic mechanism. This anti-inflammatory effect of VSL#3 was found to be mediated by *L. casei*-derived lactocepin and was proven to be of relevance in specific IBD indications *in vivo*.

The inhibitory effect of VSL#3 on IP-10 expression was revealed to be due to a single bacterial strain of the probiotic mixture, *L. casei* (VSL#3). The protective bacterial structure of *L. casei* (VSL#3) was found to be a bacterial serine protease and was identified as lactocepin type II (PrtP). Lactocepins are bacterial cell surface proteases which can also be shedded off the membrane via autoproteolysis, explaining why both, whole cell *L. casei* (VSL#3) as well as bacterial supernatants, were found to induce the inhibition of IP-10 expression in IEC. Lactocepins were first discovered in *Lactococcus lactis* and caseins were found to be the main substrate of this bacterial protease. In consequence, lactocepin-expressing bacteria are capable of using caseins for their amino acid supply (Juillard et al., 1995). There are three known types of lactocepins, PI to PIII, which show at least 95% sequence homology. The different lactocepin types are classified according to biochemical properties and according to their specificity towards casein subgroups. These subtypes can be subclassified into different groups of lactocepins (a–f) according to their preferred cleavage sites of the substrate (Holck and Naes, 1992). Apart from the respective amino acid sequence of the proteases, the specificity of lactocepins towards alpha, beta or kappa casein was found to be determined via post-translational modifications, mainly through autoproteolysis (Broadbent et al., 2002; Flambard and Juillard, 2000). Genetic engineering experiments revealed that lactocepin is still active even if large peptide sequences are excised (Bruinenberg et al., 1994). The extensive proteolytic processing of lactocepin is thought to be the reason for the very uncommon elution profile of probiotic activity after chromatographic fractionation of CM. Differently processed isoforms of lactocepin are thought to elute in separate fractions according to the respective molecular weight or charge. Consequently, several separated fractions were found to be active with regard to the inhibition of IP-10 in IEC (Figure 29). LC-MS-MS finally revealed that all active fractions contain peptide sequences that have a lactocepin type II precursor in common (Table 4), but which processed forms of this precursor are active isoforms of lactocepin remains to be elucidated.

So far, lactocepin is one of the very first probiotic structures that have been identified to exert a specific anti-inflammatory effect. To date, it is unclear why other Lactobacillus strains like L. casei BL23 and L. casei ATCC334, which both do encode lactocepins, were inactive with regard to the inhibition of IP-10 expression in IEC. A possible explanation for this finding could be that these strains do either not express lactocepin on protein level at all, or that they do not express the serine protease under the bacterial growth conditions applied in this study. In this context, the transcription of lactocepin was found to be sensitive to peptide and amino acid concentrations in the growth media (Marugg et al., 1995; Marugg et al., 1996). It could also well be that the bacterial strains that were found to be inactive with regard to the inhibition of IP-10 secretion encode a lactocepin protein that exhibits a different substrate specificity compared to the one produced by the active strains. Another possible explanation is that lactocepin is not processed into the active isoforms which are responsible for the antiinflammatory effect on IP-10 expression in the lactocepin-encoding but unactive strains. In this context, it is known that some lactocepin isoforms need to be processed into the respective active forms by a proteinase maturation protein, prtM. Furthermore, some bacterial strains were found to encode non-functional lactocepins (Cai et al., 2009).

To date, there is nothing known about the target receptor or protein on IEC which is activated or cleaved by lactocepin, resulting in the observed loss of IP-10 in IEC. Regarding the weak proteolysis of FTC-casein by CM L. casei (VSL#3) compared to the strong inhibitory effect of CM L. casei (VSL#3) on IP-10 secretion in IEC, it might be speculated that L. casei (VSL#3) produces lactocepin isoforms with high specificity for the sofar unknown receptor on IEC. Alternatively, it might well be that there are only very low amounts of lactocepin needed to induce an inhibitory cascade in IEC. The finding that CM of L. casei BL23 does not show any proteolytic activity on FTC-casein indicates that this strain does indeed not produce active forms of lactocepin.

Interestingly, the ability of a bacterial strain to reduce the level of secreted IP-10 via lactocepin might be an important survival factor, as high levels of IP-10 were found to exert antibacterial, defensin-like effects (Cole et al., 2001; Eliasson and Egesten, 2008). The evolutionary strain-specific development of lactocepin isotypes that are able to inhibit antibacterial IP-10 secretion might be an important anti-inflammatory probiotic mechanism in the context of IBD.

Inflamed intestinal tissue in IBD patients and animal models of experimental inflammation is characterized by highly elevated levels of the proinflammatory chemokine IP-10 (Banks, 2003; Hyun, 2005; Suzuki, 2007). In a healthy host, pathogen-induced secretion of IP-10 triggers recruitment of effector Th1 cells and monocytes into the mucosa, resulting in subsequent deletion of the infectious agent. In IBD, dysregulated secretion of high amounts of IP-10 in the absence of any pathogenic stimuli may lead to continuous recruitment and activation of effector cells, resulting in the loss of epithelial cell integrity and tissue damage. The present work shows for the first time that IP-10 expression in IEC correlates with inflammation in IBD. IP-10 expression was found to be strongly elevated in primary IEC from the ileum of heterozygous TNF$^{\Delta ARE/+}$ mice as well as in IEC isolated from the large intestine of IL-10-/- mice compared to healthy wildtype mice. These findings show that IEC are a major source of IP-10 within chronically inflamed intestinal tissues, suggesting a strong anti-inflammatory impact of lactocepin-reduced IP-10 secretion in the context of IBD. This hypothesis is supported by protective effects of a neutralizing anti-IP-10 antibody in an experimental study. The use of this anti-IP-10 antibody was found to be sufficient to significantly reduce colitis in IL10-/- mice. The protective effect was shown to be mediated by reduced recruitment of activated Th1 cells into the mesenteric lymph nodes as well as into the colonic mucosa of IL10-/- mice. The study also revealed that the anti-IP-10 antibody was more effective in the prevention of severe colitis than in the reduction of active disease (Hyun et al., 2005). This finding might be explained by the fact that neutralization of IP-10 does block the recruitment of proinflammatory effector cells but does not inhibit the activity of already present Th1 cells or monocytes. As a consequence of the great success of the anti-IP-10 therapy on experimental colitis in IL10-/- mice, a phase one clinical study with an anti-IP-10 antibody was performed. As the phase one trial was successfully completed, a phase II clinical study was started in 2008 but the results are not yet published (Medarex Announcement, 2008).

Since IP-10 secretion has a profound impact on health and disease, it becomes clear that the expression and secretion of this chemokine has to be tightly regulated. Consistently, IP-10 expression has been shown to be modulated at various levels. The activation of peroxisome proliferators activated receptor (PPAR) γ by PPARγ ligands was found to inhibit IFNγ-induced IP-10 expression by reducing IP-10 promoter activity in human endothelial cells (Marx et al., 2000). In addition, butyrate was revealed to inhibit IFNγ and TNF-induced IP-10 gene transcription by STAT1-independent pathways in colonic subepithelial myofibroblasts (Inatomi et al., 2005). Apart from the transcriptional regulation, processing of cytokines and chemokines at the post-transcriptional and post-translational level is suggested to be very important for their proper functionality (Proost et al., 2006; Proost, 2006) but exact mechanisms are mostly unknown. In this context, flagellin-induced-interleukin 8 expression was shown to be downregulated by the inhibition of p38 via a post-transcriptional mechanism in IEC (Yu et al., 2003). In addition, the biosynthesis of IFNγ protein was found to be selectively suppressed in Th1 cells stimulated with lead, whereas IFNγ mRNA levels were not reduced (Heo et al., 2007). Concerning IP-10, complex post-transcriptional regulatory mechanisms seem to play a very important role. Recent studies revealed that S100b-induced increase of IP-10 mRNA half-life in monocytes seems to be responsible for increased IP-10 protein production and secretion (Shanmugam et al., 2006). Furthermore, extensive post-translational proteolytic processing of IP-10 was found to take place. The truncation of four amino acids at the C-terminal end of IP-10 was found to be mediated through furin, a pro-protein convertase in keratinocytes. Furthermore, extracellular IP-10 was found to be processed C-terminally by neutrophil secreted neutrophil collagenase/matrix metallo proteinase 8, whereas Gelatinase B/matrix metallo proteinase 9 was found to degrade the chemokine completely (Van den Steen et al., 2003). The impact of C-terminal processing of IP-10 on IP-10 functionality remains to be elucidated as neither IP-10 heparin binding (Luster et al., 1995) or CXCR3 binding properties nor the direct antimicrobial effect of IP-10 is changed by this kind of posttranslational processing (Hensbergen et al., 2004). In contrast, N-terminal dipeptide cleavage by extracellular and/or intracellular proteases like CD26 dipeptidyl prolyl IV peptidase results in neutralisation of IP-10. This mechanism is very important for the termination of an inflammatory response, for example after successful elimination of pathogens from tissues. In addition to the strong reduction of its own chemotactic activity, N-terminally truncated IP-10 acts as a chemotactic antagonist for other chemokines whereas it retains its inhibitory activity on angiogenesis (Proost et al., 2001; Proost et al., 2006). A recent study reported that post-translational citrullination of IP-10 (conversion of arginine to citrullin) by a peptidylarginine deiminase resulted in similar loss of IP-10 chemotactic activity without affecting CXCR3 binding (Loos et al., 2008). In addition to post-translational processing, the functionality of IP-10 was shown to be modulated by its oligomerization status (Jabeen et al., 2008). Oligomeric but not monomeric IP-10 was found to mediate transendothelial migration of lymphocytes via its binding to endothelial cells (Campanella et al., 2006). The present work revealed for the first time that probiotic bacteria are able to induce protective post-translational regulatory mechanisms in IEC.

L. casei (VSL#3) was found to induce an IP-10 specific secretory blockade resulting in subsequent degradation of IP-10. Loss of intracellular IP-10 protein was also found to be induced by brefeldin A, a general inhibitor of protein secretion. This finding suggests that intracellular degradation of IP-10 is due to an intrinsic mechanism that is induced in IEC as a consequence of blocked IP-10 secretion. It is hypothesized that the degradation of IP-10 is a protective mechanism, preventing harmful IP-10 accumulation in IEC. Interestingly, 3-MA, an inhibitor of vesicle formation and exosome secretion (Dardalhon, 2002) specifically inhibited the secretion of IP-10. Considering the extensive post-translational processing of IP-10, the inhibition of IP-10 secretion by 3-MA might be a consequence of chemokine maturation processes in endolysosomal compartments prior to exocytosis. Endolysosomal protein maturation has been shown to play a role in the context of several other secretory proteins (Marinari, 2003; Schähs, 2008). Since 3-MA and *L. casei* (VSL#3) show exactly the same IP-10-specific inhibitory profile it is hypothesized that *L. casei* (VSL#3) mediates an inhibition of vesicular trafficking similar to 3-MA. This inhibition is thought to result in impaired secretion of IP-10, followed by degradation of the chemokine. However, proteases that are involved in the clearance of intracellular IP-10 in IEC remain to be elucidated since *L. casei* (VSL#3)-induced ubiquitination of IP-10 does not target the protein for proteasomal degradation. Furthermore, and in contrast to the finding that LPS-induced TNF is degraded in lysosomes under hypoxic conditions (Lahat, 2008), the blockade of lysosomal pathways did not rescue IP-10 protein from *L. casei* (VSL#3)-induced degradation, suggesting alternative degradation pathways to be involved.

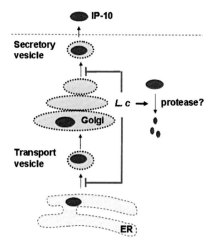

Figure 37. Hypothetic blockade of IP-10 specific vesicular secretory pathways by *L. casei* (VSL#3).
L. casei (VSL#3) is thought to block IP-10 specific vesicular secretion pathways. Blocked secretion of IP-10 induces degradation of IP-10 via unknown proteolytic pathways in IEC.

In consistence with the post-translational nature of the inhibitory mechanism induced by *L. casei* (VSL#3), IP-10 protein degradation in IEC was found to be stimuli-independent. This finding is very important for the relevance of the observed mechanism *in vivo* as IP-10 expression in IEC is triggered by various different stimuli like TNF, IFNγ and LPS. Furthermore, the finding that the inhibition of IP-10 secretion in IEC by *L. casei* (VSL#3) has a strong negative impact on the migration of activated T cells, without affecting epithelial cell viability, is promising with regard to the relevance of this mechanism *in vivo*.

In agreement with the promising *in vitro* data, VSL#3 feeding studies revealed physiological relevance of the inhibition of IP-10 expression in primary IEC, but the protective effect of VSL#3 was found to be dependent on the spatial distribution of the intestinal inflammation. VSl#3 feeding was found to be protective in the context of cecal inflammation in IL10-/- mice, whereas colonic inflammation in IL10-/- mice as well as ileal inflammation in TNF$^{\Delta ARE/+}$ mice were found to be unaffected by the probiotic treatment.

Post weaning feeding of VSL#3 or *L. casei* (VSL#3) did not exhibit protective effects on the development of ileal inflammation in TNF$^{\Delta ARE/+}$ mice. In addition, the lack of protective effects of pre-weaning feeding of VSL#3 in TNF$^{\Delta ARE/+}$ mice conclusively revealed that VSL#3 is not able to prevent the onset of ileal inflammation. The failure of probiotic therapy in the context of ileitis in TNF$^{\Delta ARE/+}$ mice could be due to a combination of several factors. First, the findings might indicate that the constant overexpression of TNF in TNF$^{\Delta ARE/+}$ mice drives such a strong Th1-cell mediated immunopathology, that the anti-inflammatory effects of probiotic therapy are too weak to confer a notable protection on the development of tissue pathology. Second, the lack of protective probiotic effects on ileitis in TNF$^{\Delta ARE/+}$ mice could also be a consequence of the disease localisation, as the ileal bacterial load is lower whereas the transit rate is higher compared to the large intestine (Savage, 1977). These conditions might account for low numbers of probiotic bacteria in the ileum. It might also be that the probiotic bacteria pass the ileum before being fully recovered from the low pH in the stomach as well as from high concentrations of bile acids and digestive enzymes in the duodenum. These stressful conditions in the upper gastrointestinal tract might be responsible for delayed protective activities of probiotic bacteria, although probiotics are characterized by the ability to survive these hostile conditions. In this context, it is also possible that the sour and highly digestive environment in the small intestine hampers probiotic functionality by the disruption of active surface components, without affecting probiotic activity and viability. The latter hypothesis could serve as an explanation for the lack of IP-10 inhibition in primary IEC of TNF$^{\Delta ARE/+}$ mice by *L. casei* (VSL#3), as the protective mechanism was found to be mediated by trypsin sensitive, cell surface bound lactocepin. Cell-surface bound lactocepin might be cleaved and inactivated by digestive proteases after the entry of *L. casei* (VSL#3) into the duodenum. It might take some time until the probiotic bacteria resynthesize and socrete the active component in protective amounts. This hypothesis suggests probiotic therapy to be more efficient in the more distal compartments of the intestine. However, the observation that VSL#3 does not reduce ileal inflammation in TNF$^{\Delta ARE/+}$ mice Is consistent with clinical studies in CD patients which mostly failed to show protective probiotic effects (Rolfe et al., 2006).

In contrast to the lack of protective effects on ileitis in TNF$^{\Delta ARE/+}$ mice, post-weaning feeding of VSL#3 was found to abrogate cecal inflammation in IL10-/- mice. Unlike the result of previous studies (Madsen et al., 2001), colonic inflammation in IL10-/- mice was not reduced by oral uptake of VSL#3. This finding suggests a strong impact of the resident colonic microbiota on the efficacy of probiotic therapy. The finding of intestinal segment specific probiotic protection in a single animal model of experimental colitis clearly shows that probiotic efficacy is dependent on the localisation of inflammation. The different response towards probiotic treatment in cecal and colonic compartments might be due to the fact that the cecum exhibits a lower inflammatory grade compared to the colon even in placebo treated IL-10-/- mice. The mildly inflamed tissue might be receptive for regulatory effects of probiotic bacteria whereas the protective pathways triggered by probiotics might be overwritten under conditions of severe inflammation. Another explanation for the different response might be the different bacterial colonisation in these two intestinal compartments enabling protective probiotic effects in the cecum but not in the colon. The intestinal microbiota has already been shown to have an enormous impact upon the onset, severity and localisation of experimental colitis, suggesting that even small differences might impact the effectiveness of probiotic treatment (Kim, 2005). To date, the reason why VSL#3 is effective in one intestinal compartment but not in the other remains speculative but detailed analysis of probiotic presence and functionality in the different intestinal compartments might resolve at least parts of these questions in the future. The severity of the colonic inflammation in VSL#3-fed IL10-/- mice likely accounts for the high plasma level of the inflammatory marker SAA, despite the fact that cecal inflammation was reduced by the probiotic therapy. This finding indicates that the reduced cecal inflammation is due to a local probiotic effect and not due to the induction of systemic regulatory mechanisms. Consistent with this hypothesis, the protective activity of VSL#3 on cecal inflammation in IL10-/- mice was found to correlate to reduced IP-10 expression in primary cecal epithelial cells. In addition, the number of CCL5 positive T cells was found to be strongly reduced in the cecum of VSL#3 treated mice (Reiff et al., 2009) indicating that the reduction of IP-10 secretion in IEC results in impaired recruitment of effector cells. Interestingly, the loss of IP-10 protein in primary IEC was found to be mediated by a post-transcriptional mechanism. This result suggests that the in vitro-observed post-translational inhibition of IP-10 secretion in IEC by L. casei (VSL#3)-derived lactocepin is indeed relevant in vivo, and contributes to the protective effect of VSL#3 in the context of IBD (Figure 38).

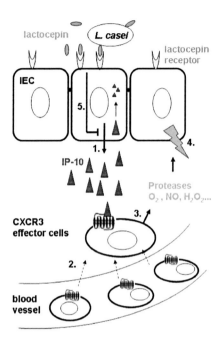

Figure 38. Postulated tissue protective effect of lactocepin *in vivo*.

Increased secretion of IP-10 from dysregulated IEC (1.) results in enhanced recruitment of proinflammatory CXCR3-bearing effector cells into the mucosal tissue along an IP-10 gradient (2.). The effector molecules produced by these cells (3.) subsequently damage the mucosal tissue cells and reduce the intestinal barrier function (4.). Cell-surface bound or secreted lactocepin from specific probiotic strains inhibits the secretion of IP-10 in IEC (5.) and thereby protects the mucosal tissue from the deleterious consequences of uncontrolled IP-10 secretion.

Future feeding experiments of IL10-/- mice including *L. casei* (VSL#3) and a lactocepin deficient isogenic mutant will finally answer the question of whether this bacterial serine protease is responsible for the observed loss of IP-10 protein in primary IEC, resulting in reduced chronic intestinal inflammation. If this holds true, the determination and purification of active lactocepin isoforms as well as the analysis of the efficacy and safety of this defined bacterial structure in experimental IBD models will be an important future task. In the best of all cases, the identification of lactocepin as an anti-inflammatory probiotic structure might lead to its successful use as a new anti-inflammatory macromolecule in pharmaceuticals or functional food preparations in the context of IBD.

LIST OF FIGURES

LIST OF TABLES

ABBREVIATIONS

APC	antigen presenting cell
ARE	adenosine uracil rich region
B.	*Bifidobacterium .....*
bd	bidest
CARD	caspase activating receptor domain
CD	Crohn´s Disease
CD4/8	complementariy determining region 4/8
cfu	colony forming units
CM	conditioned media
Cp	crossing point
DC	dendritic cell
DSS	dextran sodium sulphate
E.c. Nissle	*Escherichia coli* Nissle 1917
ELISA	enzyme linked immunosorbent assy
ER	endoplasmic reticulum
GALT	gut associated lymphoid tissue
H&E	hematoxylin&eosin
IBD	inflammatory bowel disease
IEC	intestinal epithelial cells
IEL	intraepithelial lymphocytes
IFN	interferon
Ig	immunglobulin
IκB	inhibitor κB
IKK	inhibitor κB kinase complex
IL10-/-	interleukin 10 knockout
IP-10	interferon inducible protein 10
ISRE	interferon response element
JAK	Janus kinase
IL	interleukin
LC MS MS	liquid chromatography mass spectrometry mass spectrometry
L. ...	*Lactobacillus ....*
L.c	*L. casei* (VSL#3)
LGG	*Lactobacillus rhamnosus* GG

3-MA	3-methyladenine
MAMP	microbe associated molecular pattern
MAPK	mitogen activated protein kinase
MHC	major histocompatibility complex
MIP	macrophage inflammatory protein 2
moi	multiplicity of infection
MLCK	myosin light chain kinase
MLV	murine leukemia virus
n.d.	not detectable
NFκB	nuclear factor κB
NOD	nucleotide binding oligomerisation domain
o.n.	over night
PAMP	pathogen associated molecular pattern
PCR	polymerase chain reaction
PPAR	peroxisome proliferator activated receptor
PRR	pattern recognition receptor
RT	room temperature
SAA	serum amyloid A
SC	sequence coverage
SCFA	short chain fatty acids
SCID	severe combined immunodeficiency
SDF-1	stromal derived factor 1
SDS	sodium dodecyl sulfate
SEAP	secreted embryonic alkaline phosphatase
SPF	specific pathogen free
STAT	signal transducer and activator of transcription
TCR	T cell receptor
TEAB	Tetraethylammonium bromide
TECK	thymocyte expressed chemokine
TEER	transepithelial electrical resistance
Tg	transgene
TGF	transforming growth factor
Th	T helper cell
TJ	tight junction
TNBS	trinitrobenzene sulfonic acid
TNF	tumor necrosis factor

TLR	toll like receptor
Treg	T regulatory cell
UC	ulcerative colitis
WB	Western Blot
wt	wildtype
ZO	zonula occludin

REFERENCES

Abreu, M. T., Arnold, E. T., Thomas, L. S., Gonsky, R., Zhou, Y., Hu, B. and Arditi, M. (2002). TLR4 and MD-2 expression is regulated by immune-mediated signals in human intestinal epithelial cells. *J Biol Chem* **277**, 20431-7.

Abreu, M. T., Thomas, L. S., Arnold, E. T., Lukasek, K., Michelsen, K. S. and Arditi, M. (2003). TLR signaling at the intestinal epithelial interface. *J Endotoxin Res* **9**, 322-30.

Alfaleh, K. and Bassler, D. (2008). Probiotics for prevention of necrotizing enterocolitis in preterm infants. *Cochrane Database Syst Rev*, CD005496.

Artis, D. (2008). Epithelial-cell recognition of commensal bacteria and maintenance of immune homeostasis in the gut. *Nat Rev Immunol* **8**, 411-20.

Backhed, F., Ley, R. E., Sonnenburg, J. L., Peterson, D. A. and Gordon, J. I. (2005). Host-bacterial mutualism in the human intestine. *Science* **307**, 1915-20.

Banks, C., Bateman A, Payne R, Johnson P, Sheron N. (2003). Chemokine expression in IBD. Mucosal chemokine expression is unselectively increased in both ulcerative colitis and Crohn's disease. *J Pathol* **199**, 28-35.

Barnich, N., Aguirre, J. E., Reinecker, H. C., Xavier, R. and Podolsky, D. K. (2005). Membrane recruitment of NOD2 in intestinal epithelial cells is essential for nuclear factor-{kappa}B activation in muramyl dipeptide recognition. *J Cell Biol* **170**, 21-6.

Barrett, J. C., Hansoul, S., Nicolae, D. L., Cho, J. H., Duerr, R. H., Rioux, J. D., Brant, S. R., Silverberg, M. S., Taylor, K. D., Barmada, M. M. et al. (2008). Genome-wide association defines more than 30 distinct susceptibility loci for Crohn's disease. *Nat Genet* **40**, 955-62.

Bibiloni, R., Fedorak, R. N., Tannock, G. W., Madsen, K. L., Gionchetti, P., Campieri, M., De Simone, C. and Sartor, R. B. (2005). VSL#3 probiotic-mixture induces remission in patients with active ulcerative colitis. *Am J Gastroenterol* **100**, 1539-46.

Bousvaros, A., Guandalini, S., Baldassano, R. N., Botelho, C., Evans, J., Ferry, G. D., Goldin, B., Hartigan, L., Kugathasan, S., Levy, J. et al. (2005). A randomized, double-blind trial of Lactobacillus GG versus placebo in addition to standard maintenance therapy for children with Crohn's disease. *Inflamm Bowel Dis* **11**, 833-9.

Brandtzaeg, P., Carlsen, H. S. and Halstensen, T. S. (2006). The B-cell system in inflammatory bowel disease. *Adv Exp Med Biol* **579**, 149-67.

Broadbent, J. R., Barnes, M., Brennand, C., Strickland, M., Houck, K., Johnson, M. E. and Steele, J. L. (2002). Contribution of Lactococcus lactis cell envelope proteinase specificity to peptide accumulation and bitterness in reduced-fat Cheddar cheese. *Appl Environ Microbiol* **68**, 1778-85.

Bruinenberg, P. G., Doesburg, P., Alting, A. C., Exterkate, F. A., de Vos, W. M. and Siezen, R. J. (1994). Evidence for a large dispensable segment in the subtilisin-like catalytic domain of the Lactococcus lactis cell-envelope proteinase. *Protein Eng* **7**, 991-6.

Caballero-Franco, C., Keller, K., De Simone, C. and Chadee, K. (2007). The VSL#3 probiotic formula induces mucin gene expression and secretion in colonic epithelial cells. *Am J Physiol Gastrointest Liver Physiol* **292**, G315-22.

Cai, H., Thompson, R., Budinich, M. F., Broadbent, J. R. and Steele, J. L. (2009). Genome Sequence and Comparative Genome Analysis of Lactobacillus casei: Insights into Their Niche-Associated Evolution
10.1093/gbe/evp019. *Genome Biol Evol* **2009**, 239-257.

Campanella, G. S., Grimm, J., Manice, L. A., Colvin, R. A., Medoff, B. D., Wojtkiewicz, G. R., Weissleder, R. and Luster, A. D. (2006). Oligomerization of CXCL10 is necessary for endothelial cell presentation and in vivo activity. *J Immunol* **177**, 6991-8.

Chermesh, I., Tamir, A., Reshef, R., Chowers, Y., Suissa, A., Katz, D., Gelber, M., Halpern, Z., Bengmark, S. and Eliakim, R. (2007). Failure of Synbiotic 2000 to prevent postoperative recurrence of Crohn's disease. *Dig Dis Sci* **52**, 385-9.

Cho, J. H. (2008). The genetics and immunopathogenesis of inflammatory bowel disease. *Nat Rev Immunol* **8**, 458-66.

Cole, A. M., Ganz, T., Liese, A. M., Burdick, M. D., Liu, L. and Strieter, R. M. (2001). Cutting edge: IFN-inducible ELR- CXC chemokines display defensin-like antimicrobial activity. *J Immunol* **167**, 623-7.

Conte MP, S. S., Zamboni I, Penta M, Chiarini F, Seganti L, Osborn J, Falconieri P, Borrelli O, Cucchiara S. (2006). Gut-associated bacterial microbiota in paediatric patients with inflammatory bowel disease. *Gut* **55**, 1760-7.

Cui, H. H., Chen, C. L., Wang, J. D., Yang, Y. J., Cun, Y., Wu, J. B., Liu, Y. H., Dan, H. L., Jian, Y. T. and Chen, X. Q. (2004). Effects of probiotic on intestinal mucosa of patients with ulcerative colitis. *World J Gastroenterol* **10**, 1521-5.

Cummings, J. H. and Macfarlane, G. T. (1997). Role of intestinal bacteria in nutrient metabolism. *JPEN J Parenter Enteral Nutr* **21**, 357-65.

Dardalhon, V., Géminard C, Reggion H, Vidal M, Sainte-Marie J. (2002). Fractionation analysis of the endosomal compartment during rat reticulocyte maturation. *Cell Biol Int* **26**, 669-78.

Di Giacinto, C., Marinaro, M., Sanchez, M., Strober, W. and Boirivant, M. (2005). Probiotics ameliorate recurrent Th1-mediated murine colitis by inducing IL-10 and IL-10-dependent TGF-beta-bearing regulatory cells. *J Immunol* **174**, 3237-46.

Doganci, A., Neurath, M. F. and Finotto, S. (2005). Mucosal immunoregulation: transcription factors as possible therapeutic targets. *Curr Drug Targets Inflamm Allergy* **4**, 565-75.

Eckburg, P. B., Bik, E. M., Bernstein, C. N., Purdom, E., Dethlefsen, L., Sargent, M., Gill, S. R., Nelson, K. E. and Relman, D. A. (2005). Diversity of the human intestinal microbial flora. *Science* **308**, 1635-8.

Eliasson, M. and Egesten, A. (2008). Antibacterial chemokines--actors in both innate and adaptive immunity. *Contrib Microbiol* **15**, 101-17.

Fitzpatrick, L. R., Hertzog, K. L., Quatse, A. L., Koltun, W. A., Small, J. S. and Vrana, K. (2007). Effects of the probiotic formulation VSL#3 on colitis in weanling rats. *J Pediatr Gastroenterol Nutr* **44**, 561-70.

Flambard, B. and Juillard, V. (2000). The autoproteolysis of Lactococcus lactis lactocepin III affects its specificity towards beta-casein. *Appl Environ Microbiol* **66**, 5134-40.

Foligne, B., Zoumpopoulou, G., Dewulf, J., Ben Younes, A., Chareyre, F., Sirard, J. C., Pot, B. and Grangette, C. (2007). A key role of dendritic cells in probiotic functionality. *PLoS ONE* **2**, e313.

Frank, D. N., St Amand, A. L., Feldman, R. A., Boedeker, E. C., Harpaz, N. and Pace, N. R. (2007). Molecular-phylogenetic characterization of microbial community imbalances in human inflammatory bowel diseases. *Proc Natl Acad Sci U S A* **104**, 13780-5.

Fujino, S., Andoh, A., Bamba, S., Ogawa, A., Hata, K., Araki, Y., Bamba, T. and Fujiyama, Y. (2003). Increased expression of interleukin 17 in inflammatory bowel disease. *Gut* **52**, 65-70.

Fuller, R. (1989). Probiotics in man and animals. *J Appl Bacteriol* **66**, 365-78.

Furrie, E., Macfarlane, S., Kennedy, A., Cummings, J. H., Walsh, S. V., O'Neil D, A. and Macfarlane, G. T. (2005). Synbiotic therapy (Bifidobacterium longum/Synergy 1) initiates resolution of inflammation in patients with active ulcerative colitis: a randomised controlled pilot trial. *Gut* **54**, 242-9.

Gassull, M. A. (2006). Review article: the intestinal lumen as a therapeutic target in inflammatory bowel disease. *Aliment Pharmacol Ther* **24 Suppl 3**, 90-5.

Gaudier, E., Michel, C., Segain, J. P., Cherbut, C. and Hoebler, C. (2005). The VSL# 3 probiotic mixture modifies microflora but does not heal chronic dextran-sodium sulfate-induced colitis or reinforce the mucus barrier in mice. *J Nutr* **135**, 2753-61.

Ginzburg, L., Colp, R. and Sussman, M. (1939). Ileocolostomy with Exclusion. *Ann Surg* **110**, 648-58.

Gionchetti, P., Rizzello, F., Helwig, U., Venturi, A., Lammers, K. M., Brigidi, P., Vitali, B., Poggioli, G., Miglioli, M. and Campieri, M. (2003). Prophylaxis of pouchitis onset with probiotic therapy: a double-blind, placebo-controlled trial. *Gastroenterology* **124**, 1202-9.

Gionchetti, P., Rizzello, F., Morselli, C., Poggioli, G., Tambasco, R., Calabrese, C., Brigidi, P., Vitali, B., Straforini, G. and Campieri, M. (2007). High-dose probiotics for the treatment of active pouchitis. *Dis Colon Rectum* **50**, 2075-82; discussion 2082-4.

Gionchetti, P., Rizzello, F., Venturi, A., Brigidi, P., Matteuzzi, D., Bazzocchi, G., Poggioli, G., Miglioli, M. and Campieri, M. (2000). Oral bacteriotherapy as maintenance treatment in patients with chronic pouchitis: a double-blind, placebo-controlled trial. *Gastroenterology* **119**, 305-9.

Gordon, H. A. (1959). Morphological and physiological characterization of germfree life. *Ann N Y Acad Sci* **78**, 208-20.

Haarman, M. and Knol, J. (2006). Quantitative real-time PCR analysis of fecal Lactobacillus species in infants receiving a prebiotic infant formula. *Appl Environ Microbiol* **72**, 2359-65.

Haller, D. (2006). Intestinal epithelial cell signalling and host-derived negative regulators under chronic inflammation: to be or not to be activated determines the balance towards commensal bacteria. *Neurogastroenterol Motil* **18**, 184-99.

Haller, D. and Jobin, C. (2004). Interaction between resident luminal bacteria and the host: can a healthy relationship turn sour? *J Pediatr Gastroenterol Nutr* **38**, 123-36.

Haller, D., Russo, M. P., Sartor, R. B. and Jobin, C. (2002). IKK beta and phosphatidylinositol 3-kinase/Akt participate in non-pathogenic Gram-negative enteric bacteria-induced RelA phosphorylation and NF-kappa B activation in both primary and intestinal epithelial cell lines. *J Biol Chem* **277**, 38168-78.

Hanauer, S. B. (2006). Inflammatory bowel disease: epidemiology, pathogenesis, and therapeutic opportunities. *Inflamm Bowel Dis* **12 Suppl 1**, S3-9.

Harmsen, H. J., Wildeboer-Veloo, A. C., Raangs, G. C., Wagendorp, A. A., Klijn, N., Bindels, J. G. and Welling, G. W. (2000). Analysis of intestinal flora development in breast-fed and formula-fed infants by using molecular identification and detection methods. *J Pediatr Gastroenterol Nutr* **30**, 61-7.

Hart, A. L., Lammers, K., Brigidi, P., Vitali, B., Rizzello, F., Gionchetti, P., Campieri, M., Kamm, M. A., Knight, S. C. and Stagg, A. J. (2004). Modulation of human dendritic cell phenotype and function by probiotic bacteria. *Gut* **53**, 1602-9.

Heazlewood, C. K., Cook, M. C., Eri, R., Price, G. R., Tauro, S. B., Taupin, D., Thornton, D. J., Png, C. W., Crockford, T. L., Cornall, R. J. et al. (2008). Aberrant mucin assembly in mice causes endoplasmic reticulum stress and spontaneous inflammation resembling ulcerative colitis. *PLoS Med* **5**, e54.

Hensbergen, P. J., Verzijl, D., Balog, C. I., Dijkman, R., van der Schors, R. C., van der Raaij-Helmer, E. M., van der Plas, M. J., Leurs, R., Deelder, A. M., Smit, M. J. et al. (2004). Furin is a chemokine-modifying enzyme: in vitro and in vivo processing of CXCL10 generates a C-terminally truncated chemokine retaining full activity. *J Biol Chem* **279**, 13402-11.

Heo, Y., Mondal, T. K., Gao, D., Kasten-Jolly, J., Kishikawa, H. and Lawrence, D. A. (2007). Posttranscriptional inhibition of interferon-gamma production by lead. *Toxicol Sci* **96**, 92-100.

Hoarau, C., Martin, L., Faugaret, D., Baron, C., Dauba, A., Aubert-Jacquin, C., Velge-Roussel, F. and Lebranchu, Y. (2008). Supernatant from bifidobacterium differentially modulates transduction signaling pathways for biological functions of human dendritic cells. *PLoS ONE* **3**, e2753.

Hoffmann, M., Rath, E., Holzlwimmer, G., Quintanilla-Martinez, L., Loach, D., Tannock, G. and Haller, D. (2008). Lactobacillus reuteri 100-23 transiently activates intestinal epithelial cells of mice that have a complex microbiota during early stages of colonization. *J Nutr* **138**, 1684-91.

Holck, A. and Naes, H. (1992). Cloning, sequencing and expression of the gene encoding the cell-envelope-associated proteinase from Lactobacillus paracasei subsp. paracasei NCDO 151. *J Gen Microbiol* **138**, 1353-64.

Holdeman, L. V., Good, I. J. and Moore, W. E. (1976). Human fecal flora: variation in bacterial composition within individuals and a possible effect of emotional stress. *Appl Environ Microbiol* **31**, 359-75.

Hooper, L. V., Wong, M. H., Thelin, A., Hansson, L., Falk, P. G. and Gordon, J. I. (2001). Molecular analysis of commensal host-microbial relationships in the intestine. *Science* **291**, 881-4.

Hormannsperger, G., Clavel, T., Hoffmann, M., Reiff, C., Kelly, D., Loh, G., Blaut, M., Holzlwimmer, G., Laschinger, M. and Haller, D. (2009). Post-translational inhibition of IP-10 secretion in IEC by probiotic bacteria: impact on chronic inflammation. *PLoS ONE* **4**, e4365.

Hwang, J. M. and Varma, M. G. (2008). Surgery for inflammatory bowel disease. *World J Gastroenterol* **14**, 2678-90.

Hyun, J., Lee G, Brown JB, Grimm GR, Tang Y, Mittal N, Dirisina R, Zhang Z, FryerJP, Weinstock JV, Luster AD, Barrett TA. (2005). Anti-interferon-inducible chemokine, CXCL10, reduces colitis by impairing T helper-1 induction and recruitment in mice. *Inflamm Bowel Dis* **11**, 799-805.

Hyun, J. G., Lee, G., Brown, J. B., Grimm, G. R., Tang, Y., Mittal, N., Dirisina, R., Zhang, Z., Fryer, J. P., Weinstock, J. V. et al. (2005). Anti-interferon-inducible chemokine, CXCL10, reduces colitis by impairing T helper-1 induction and recruitment in mice. *Inflamm Bowel Dis* **11**, 799-805.

Inatomi, O., Andoh, A., Kitamura, K., Yasui, H., Zhang, Z. and Fujiyama, Y. (2005). Butyrate blocks interferon-gamma-inducible protein-10 release in human intestinal subepithelial myofibroblasts. *J Gastroenterol* **40**, 483-9.

Ishikawa, H., Akedo, I., Umesaki, Y., Tanaka, R., Imaoka, A. and Otani, T. (2003). Randomized controlled trial of the effect of bifidobacteria-fermented milk on ulcerative colitis. *J Am Coll Nutr* **22**, 56-63.

Jabeen, T., Leonard, P., Jamaluddin, H. and Acharya, K. R. (2008). Structure of mouse IP-10, a chemokine. *Acta Crystallogr D Biol Crystallogr* **64**, 611-9.

Jijon, H., Backer, J., Diaz, H., Yeung, H., Thiel, D., McKaigney, C., De Simone, C. and Madsen, K. (2004). DNA from probiotic bacteria modulates murine and human epithelial and immune function. *Gastroenterology* **126**, 1358-73.

Juillard, V., Laan, H., Kunji, E. R., Jeronimus-Stratingh, C. M., Bruins, A. P. and Konings, W. N. (1995). The extracellular PI-type proteinase of Lactococcus lactis hydrolyzes beta-casein into more than one hundred different oligopeptides. *J Bacteriol* **177**, 3472-8.

Kalliomaki, M., Collado, M. C., Salminen, S. and Isolauri, E. (2008). Early differences in fecal microbiota composition in children may predict overweight. *Am J Clin Nutr* **87**, 534-8.

Kamada, N., Inoue, N., Hisamatsu, T., Okamoto, S., Matsuoka, K., Sato, T., Chinen, H., Hong, K. S., Yamada, T., Suzuki, Y. et al. (2005). Nonpathogenic Escherichia coli strain Nissle1917 prevents murine acute and chronic colitis. *Inflamm Bowel Dis* **11**, 455-63.

Kamada, N., Maeda, K., Inoue, N., Hisamatsu, T., Okamoto, S., Hong, K. S., Yamada, T., Watanabe, N., Tsuchimoto, K., Ogata, H. et al. (2008). Nonpathogenic Escherichia coli strain Nissle 1917 inhibits signal transduction in intestinal epithelial cells. *Infect Immun* **76**, 214-20.

Katakura, K., Lee J, Rachmilewitz D, Li G, Eckmann L, Raz E. (2005). Toll-like receptor 9-induced type I IFN protects mice from experimental colitis. *J Clin Invest* **115**, 695-702.

Kato, K., Mizuno, S., Umesaki, Y., Ishii, Y., Sugitani, M., Imaoka, A., Otsuka, M., Hasunuma, O., Kurihara, R., Iwasaki, A. et al. (2004). Randomized placebo-controlled trial assessing the effect of bifidobacteria-fermented milk on active ulcerative colitis. *Aliment Pharmacol Ther* **20**, 1133-41.

Kim, S., Tonkonogy SL, Albright CA, Tsang J, Balish EJ, Braun J, Huycke MM, Sartor RB. (2005). Variable phenotypes of enterocolitis in interleukin 10-deficient mice monoassociated with two different commensal bacteria. *Gastroenterology* **128**, 891-906.

Kokesova, A., Frolova, L., Kverka, M., Sokol, D., Rossmann, P., Bartova, J. and Tlaskalova-Hogenova, H. (2006). Oral administration of probiotic bacteria (E. coli Nissle, E. coli O83, Lactobacillus casei) influences the severity of dextran sodium sulfate-induced colitis in BALB/c mice. *Folia Microbiol (Praha)* **51**, 478-84.

Kontoyiannis, D., Pasparakis, M., Pizarro, T. T., Cominelli, F. and Kollias, G. (1999). Impaired on/off regulation of TNF biosynthesis in mice lacking TNF AU-rich elements: implications for joint and gut-associated immunopathologies. *Immunity* **10**, 387-98.

Kraus, T. A., Toy, L., Chan, L., Childs, J., Cheifetz, A. and Mayer, L. (2004). Failure to induce oral tolerance in Crohn's and ulcerative colitis patients: possible genetic risk. *Ann N Y Acad Sci* **1029**, 225-38.

Kruis, W., Fric, P., Pokrotnieks, J., Lukas, M., Fixa, B., Kascak, M., Kamm, M. A., Weismueller, J., Beglinger, C., Stolte, M. et al. (2004). Maintaining remission of ulcerative colitis with the probiotic Escherichia coli Nissle 1917 is as effective as with standard mesalazine. *Gut* **53**, 1617-23.

Kruis, W., Schutz, E., Fric, P., Fixa, B., Judmaier, G. and Stolte, M. (1997). Double-blind comparison of an oral Escherichia coli preparation and mesalazine in maintaining remission of ulcerative colitis. *Aliment Pharmacol Ther* **11**, 853-8.

Kuisma, J., Mentula, S., Jarvinen, H., Kahri, A., Saxelin, M. and Farkkila, M. (2003). Effect of Lactobacillus rhamnosus GG on ileal pouch inflammation and microbial flora. *Aliment Pharmacol Ther* **17**, 509-15.

Lahat, N., Rahat MA, Kinarty A, Weiss-Cerem L, Pinchevski S, Bittermann H. (2008). Hypoxia enhances lysosomal TNF(alpha) degradation in mouse peritoneal makrophages. *Am J Physiol Cell Physiol* **In press**.

Lammers, K. M., Vergopoulos, A., Babel, N., Gionchetti, P., Rizzello, F., Morselli, C., Caramelli, E., Fiorentino, M., d'Errico, A., Volk, H. D. et al. (2005). Probiotic therapy in the prevention of pouchitis onset: decreased interleukin-1beta, interleukin-8, and interferon-gamma gene expression. *Inflamm Bowel Dis* **11**, 447-54.

Levy, J. (2000). The effects of antibiotic use on gastrointestinal function. *Am J Gastroenterol* **95**, S8-10.

Lilly, D. M. and Stillwell, R. H. (1965). Probiotics: Growth-Promoting Factors Produced by Microorganisms. *Science* **147**, 747-8.

Loftus, E. (2004). Clinical epidemiology of inflammatory bowel disease: Incidence, prevalence, and environmental influences. *Gasteroenterology* **126**, 1504-17.

Loos, T., Mortier, A., Gouwy, M., Ronsse, I., Put, W., Lenaerts, J. P., Van Damme, J. and Proost, P. (2008). Citrullination of CXCL10 and CXCL11 by peptidylarginine deiminase: a naturally occurring posttranslational modification of chemokines and new dimension of immunoregulation. *Blood* **112**, 2648-56.

Luster, A. D., Greenberg, S. M. and Leder, P. (1995). The IP-10 chemokine binds to a specific cell surface heparan sulfate shared with platelet factor 4 and inhibits endothelial cell proliferation. *J Exp Med* **182**, 219-31.

Mackie, R. I., Sghir, A. and Gaskins, H. R. (1999). Developmental microbial ecology of the neonatal gastrointestinal tract. *Am J Clin Nutr* **69**, 1035S-1045S.

Madsen, K., Cornish, A., Soper, P., McKaigney, C., Jijon, H., Yachimec, C., Doyle, J., Jewell, L. and De Simone, C. (2001). Probiotic bacteria enhance murine and human intestinal epithelial barrier function. *Gastroenterology* **121**, 580-91.

Malchow, H. A. (1997). Crohn's disease and Escherichia coli. A new approach in therapy to maintain remission of colonic Crohn's disease? *J Clin Gastroenterol* **25**, 653-8.

Marchesi, J. R., Holmes, E., Khan, F., Kochhar, S., Scanlan, P., Shanahan, F., Wilson, I. D. and Wang, Y. (2007). Rapid and noninvasive metabonomic characterization of inflammatory bowel disease. *J Proteome Res* **6**, 546-51.

Marinari, U., Nitti M, Pronzato MA, Domenicotti C. (2003). Role of PKC-dependent pathways in HNE-induced cell protein transport and secretion. *Mol Aspects Med.* **24**, 205-11.

Marteau, P., Lemann, M., Seksik, P., Laharie, D., Colombel, J. F., Bouhnik, Y., Cadiot, G., Soule, J. C., Bourreille, A., Metman, E. et al. (2006). Ineffectiveness of Lactobacillus johnsonii LA1 for prophylaxis of postoperative recurrence in Crohn's disease: a randomised, double blind, placebo controlled GETAID trial. *Gut* **55**, 842-7.

Marugg, J. D., Meijer, W., van Kranenburg, R., Laverman, P., Bruinenberg, P. G. and de Vos, W. M. (1995). Medium-dependent regulation of proteinase gene expression in Lactococcus lactis: control of transcription initiation by specific dipeptides. *J Bacteriol* **177**, 2982-9.

Marugg, J. D., van Kranenburg, R., Laverman, P., Rutten, G. A. and de Vos, W. M. (1996). Identical transcriptional control of the divergently transcribed prtP and prtM genes that are required for proteinase production in lactococcus lactis SK11. *J Bacteriol* **178**, 1525-31.

Marx, N., Mach, F., Sauty, A., Leung, J. H., Sarafi, M. N., Ransohoff, R. M., Libby, P., Plutzky, J. and Luster, A. D. (2000). Peroxisome proliferator-activated receptor-gamma activators inhibit IFN-gamma-induced expression of the T cell-active CXC chemokines IP-10, Mig, and I-TAC in human endothelial cells. *J Immunol* **164**, 6503-8.

Mazmanian, S. K., Liu, C. H., Tzianabos, A. O. and Kasper, D. L. (2005). An immunomodulatory molecule of symbiotic bacteria directs maturation of the host immune system. *Cell* **122**, 107-18.

McGovern, D. P., Taylor, K. D., Landers, C., Derkowski, C., Dutridge, D., Dubinsky, M., Ippoliti, A., Vasiliauskas, E., Mei, L., Mengesha, E. et al. (2009). MAGI2 genetic variation and inflammatory bowel disease. *Inflamm Bowel Dis* **15**, 75-83.

Medzhitov, R. (2001). Toll-like receptors and innate immunity. *Nat Rev Immunol* **1**, 135-45.

Medzhitov, R. and Janeway, C., Jr. (2000). Innate immune recognition: mechanisms and pathways. *Immunol Rev* **173**, 89-97.

Messlik, A., Schmechel, S., Kisling, S., Bereswill, S., Heimesaat, M. M., Fischer, A., Gobel, U. and Haller, D. (2009). Loss of Toll-like receptor 2 and 4 leads to differential induction of endoplasmic reticulum stress and proapoptotic responses in the intestinal epithelium under conditions of chronic inflammation. *J Proteome Res* **8**, 4406-17.

Mimura, T., Rizzello, F., Helwig, U., Poggioli, G., Schreiber, S., Talbot, I. C., Nicholls, R. J., Gionchetti, P., Campieri, M. and Kamm, M. A. (2004). Once daily high dose probiotic therapy (VSL#3) for maintaining remission in recurrent or refractory pouchitis. *Gut* **53**, 108-14.

Moehle, C., Ackermann, N., Langmann, T., Aslanidis, C., Kel, A., Kel-Margoulis, O., Schmitz-Madry, A., Zahn, A., Stremmel, W. and Schmitz, G. (2006). Aberrant intestinal expression and allelic variants of mucin genes associated with inflammatory bowel disease. *J Mol Med* **84**, 1055-66.

Moore, W. E. and Holdeman, L. V. (1974). Human fecal flora: the normal flora of 20 Japanese-Hawaiians. *Appl Microbiol* **27**, 961-79.

Neutra, M. R., Mantis, N. J. and Kraehenbuhl, J. P. (2001). Collaboration of epithelial cells with organized mucosal lymphoid tissues. *Nat Immunol* **2**, 1004-9.

Niedzielin, K., Kordecki, H. and Birkenfeld, B. (2001). A controlled, double-blind, randomized study on the efficacy of Lactobacillus plantarum 299V in patients with irritable bowel syndrome. *Eur J Gastroenterol Hepatol* **13**, 1143-7.

Niess, J. H., Brand, S., Gu, X., Landsman, L., Jung, S., McCormick, B. A., Vyas, J. M., Boes, M., Ploegh, H. L., Fox, J. G. et al. (2005). CX3CR1-mediated dendritic cell access to the intestinal lumen and bacterial clearance. *Science* **307**, 254-8.

Nishimura, M., Kuboi, Y., Muramoto, K., Kawano, T. and Imai, T. (2009). Chemokines as novel therapeutic targets for inflammatory bowel disease. *Ann N Y Acad Sci* **1173**, 350-6.

Otte, J. M., Cario, E. and Podolsky, D. K. (2004). Mechanisms of cross hyporesponsiveness to Toll-like receptor bacterial ligands in intestinal epithelial cells. *Gastroenterology* **126**, 1054-70.

Peran, L., Camuesco, D., Comalada, M., Nieto, A., Concha, A., Adrio, J. L., Olivares, M., Xaus, J., Zarzuelo, A. and Galvez, J. (2006). Lactobacillus fermentum, a probiotic capable to release glutathione, prevents colonic inflammation in the TNBS model of rat colitis. *Int J Colorectal Dis* **21**, 737-46.

Petrof, E. O., Kojima, K., Ropeleski, M. J., Musch, M. W., Tao, Y., De Simone, C. and Chang, E. B. (2004). Probiotics inhibit nuclear factor-kappaB and induce heat shock proteins in colonic epithelial cells through proteasome inhibition. *Gastroenterology* **127**, 1474-87.

Pfaffl, M. W. (2001). A new mathematical model for relative quantification in real-time RT-PCR. *Nucleic Acids Res* **29**, e45.

Prantera, C., Scribano, M. L., Falasco, G., Andreoli, A. and Luzi, C. (2002). Ineffectiveness of probiotics in preventing recurrence after curative resection for Crohn's disease: a randomised controlled trial with Lactobacillus GG. *Gut* **51**, 405-9.

Pronio, A., Montesani, C., Butteroni, C., Vecchione, S., Mumolo, G., Vestri, A., Vitolo, D. and Boirivant, M. (2008). Probiotic administration in patients with ileal pouch-anal anastomosis for ulcerative colitis is associated with expansion of mucosal regulatory cells. *Inflamm Bowel Dis* **14**, 662-8.

Proost, P., Schutyser, E., Menten, P., Struyf, S., Wuyts, A., Opdenakker, G., Detheux, M., Parmentier, M., Durinx, C., Lambeir, A. M. et al. (2001). Amino-terminal truncation of CXCR3 agonists impairs receptor signaling and lymphocyte chemotaxis, while preserving antiangiogenic properties. *Blood* **98**, 3554-61.

Proost, P., Struyf, S. and Van Damme, J. (2006). Natural post-translational modifications of chemokines. *Biochem Soc Trans* **34**, 997-1001.

Proost, P., Struyf S, Van Damme J. (2006). Natural post-translational modifications of chemokines. *Biochem Soc Trans.* **34**, 997-1001.

Rachmilewitz, D., Katakura, K., Karmeli, F., Hayashi, T., Reinus, C., Rudensky, B., Akira, S., Takeda, K., Lee, J., Takabayashi, K. et al. (2004). Toll-like receptor 9 signaling mediates the anti-inflammatory effects of probiotics in murine experimental colitis. *Gastroenterology* **126**, 520-8.

Ramasundara, M., Leach, S. T., Lemberg, D. A. and Day, A. S. (2009). Defensins and inflammation: the role of defensins in inflammatory bowel disease. *J Gastroenterol Hepatol* **24**, 202-8.

Reiff, C., Delday, M., Rucklidge, G., Reid, M., Duncan, G., Wohlgemuth, S., Hormannsperger, G., Loh, G., Blaut, M., Collie-Duguid, E. et al. (2009). Balancing inflammatory, lipid, and xenobiotic signaling pathways by VSL#3, a biotherapeutic agent, in the treatment of inflammatory bowel disease. *Inflamm Bowel Dis*.

Rembacken, B. J., Snelling, A. M., Hawkey, P. M., Chalmers, D. M. and Axon, A. T. (1999). Non-pathogenic Escherichia coli versus mesalazine for the treatment of ulcerative colitis: a randomised trial. *Lancet* **354**, 635-9.

Roediger, W. E. (1980). The colonic epithelium in ulcerative colitis: an energy-deficiency disease? *Lancet* **2**, 712-5.

Rolfe, V. E., Fortun, P. J., Hawkey, C. J. and Bath-Hextall, F. (2006). Probiotics for maintenance of remission in Crohn's disease. *Cochrane Database Syst Rev*, CD004826.

Rollins, B. J. (1997). Chemokines. *Blood* **90**, 909-28.

Roselli, M., Finamore, A., Nuccitelli, S., Carnevali, P., Brigidi, P., Vitali, B., Nobili, F., Rami, R., Garaguso, I. and Mengheri, E. (2009). Prevention of TNBS-induced colitis by different Lactobacillus and Bifidobacterium strains is associated with an expansion of gammadeltaT and regulatory T cells of intestinal intraepithelial lymphocytes. *Inflamm Bowel Dis* **15**, 1526-36.

Ruemmele, F. M., Gurbindo, C., Mansour, A. M., Marchand, R., Levy, E. and Seidman, E. G. (1998). Effects of interferon gamma on growth, apoptosis, and MHC class II expression of immature rat intestinal crypt (IEC-6) cells. *J Cell Physiol* **176**, 120-6.

Ruiz, P., Shkoda A, Kim SC, Sartor RB, Haller D. (2005). IL-10 gene deficient mice lack TGFß/Smad signalling and fail to inhibit proinflammatory gene expression in intestinal epithelial cells after the colonisation with colitogenic Enterococcus faecalis. *J Immunol* **174**, 2990-9.

Ruiz, P. A., Shkoda, A., Kim, S. C., Sartor, R. B. and Haller, D. (2005). IL-10 gene-deficient mice lack TGF-beta/Smad signaling and fail to inhibit proinflammatory gene expression in intestinal epithelial cells after the colonization with colitogenic Enterococcus faecalis. *J Immunol* **174**, 2990-9.

Salzman, N. H., Underwood, M. A. and Bevins, C. L. (2007). Paneth cells, defensins, and the commensal microbiota: a hypothesis on intimate interplay at the intestinal mucosa. *Semin Immunol* **19**, 70-83.

Sartor, R. B. (2008). Microbial influences in inflammatory bowel diseases. *Gastroenterology* **134**, 577-94.

Sartor, R. B. and Muehlbauer, M. (2007). Microbial host interactions in IBD: implications for pathogenesis and therapy. *Curr Gastroenterol Rep* **9**, 497-507.

Savage, D. C. (1977). Microbial ecology of the gastrointestinal tract. *Annu Rev Microbiol* **31**, 107-33.

Sazawal, S., Hiremath, G., Dhingra, U., Malik, P., Deb, S. and Black, R. E. (2006). Efficacy of probiotics in prevention of acute diarrhoea: a meta-analysis of masked, randomised, placebo-controlled trials. *Lancet Infect Dis* **6**, 374-82.

Schähs, P., Weidinger P, Probst OC, Svoboda B, Stadlmann J, Heug H, Waerner T, Mach L. (2008). Cellular repressor of E1A-stimulated genes is a bona fide lysosomal protein which undergoes proteolytic maturation during its biosynthesis. *Exp Cell Res* **in press**.

Scheerens, H., Hessel, E., de Waal-Malefyt, R., Leach, M. W. and Rennick, D. (2001). Characterization of chemokines and chemokine receptors in two murine models of inflammatory bowel disease: IL-10-/- mice and Rag-2-/- mice reconstituted with CD4+CD45RBhigh T cells. *Eur J Immunol* **31**, 1465-74.

Schlee, M., Harder, J., Koten, B., Stange, E. F., Wehkamp, J. and Fellermann, K. (2008). Probiotic lactobacilli and VSL#3 induce enterocyte beta-defensin 2. *Clin Exp Immunol* **151**, 528-35.

Schlee, M., Wehkamp, J., Altenhoefer, A., Oelschlaeger, T. A., Stange, E. F. and Fellermann, K. (2007). Induction of human beta-defensin 2 by the probiotic Escherichia coli Nissle 1917 is mediated through flagellin. *Infect Immun* **75**, 2399-407.

Schmoeger, R. (1960). [Bactisubtil for the treatment of infant nutrition disorders and for dysbacteria caused by enteral antibiotic administration.]. *Munch Med Wochenschr* **102**, 1213-7.

Schultz, M., Scholmerich, J. and Rath, H. C. (2003). Rationale for probiotic and antibiotic treatment strategies in inflammatory bowel diseases. *Dig Dis* **21**, 105-28.

Schultz, M., Strauch, U. G., Linde, H. J., Watzl, S., Obermeier, F., Gottl, C., Dunger, N., Grunwald, N., Scholmerich, J. and Rath, H. C. (2004a). Preventive effects of Escherichia coli strain Nissle 1917 on acute and chronic intestinal inflammation in two different murine models of colitis. *Clin Diagn Lab Immunol* **11**, 372-8.

Schultz, M., Timmer, A., Herfarth, H. H., Sartor, R. B., Vanderhoof, J. A. and Rath, H. C. (2004b). Lactobacillus GG in inducing and maintaining remission of Crohn's disease. *BMC Gastroenterol* **4**, 5.

Schulzke, J. D., Ploeger, S., Amasheh, M., Fromm, A., Zeissig, S., Troeger, H., Richter, J., Bojarski, C., Schumann, M. and Fromm, M. (2009). Epithelial tight junctions in intestinal inflammation. *Ann N Y Acad Sci* **1165**, 294-300.

Sepp, E., Julge, K., Mikelsaar, M. and Bjorksten, B. (2005). Intestinal microbiota and immunoglobulin E responses in 5-year-old Estonian children *Clin Exp Allergy* **35**, 1141-6.

Shanmugam, N., Ransohoff, R. M. and Natarajan, R. (2006). Interferon-gamma-inducible protein (IP)-10 mRNA stabilized by RNA-binding proteins in monocytes treated with S100b. *J Biol Chem* **281**, 31212-21.

Shibolet, O., Karmeli, F., Eliakim, R., Swennen, E., Brigidi, P., Gionchetti, P., Campieri, M., Morgenstern, S. and Rachmilewitz, D. (2002). Variable response to probiotics in two models of experimental colitis in rats. *Inflamm Bowel Dis* **8**, 399-406.

Shkoda, A., Werner, T., Daniel, H., Gunckel, M., Rogler, G. and Haller, D. (2007). Differential protein expression profile in the intestinal epithelium from patients with inflammatory bowel disease. *J Proteome Res* **6**, 1114-25.

Singh, U. P., Venkataraman, C., Singh, R. and Lillard, J. W., Jr. (2007). CXCR3 axis: role in inflammatory bowel disease and its therapeutic implication. *Endocr Metab Immune Disord Drug Targets* **7**, 111-23.

Sokol, H., Pigneur, B., Watterlot, L., Lakhdari, O., Bermudez-Humaran, L. G., Gratadoux, J. J., Blugeon, S., Bridonneau, C., Furet, J. P., Corthier, G. et al. (2008). Faecalibacterium prausnitzii is an anti-inflammatory commensal bacterium identified by gut microbiota analysis of Crohn disease patients. *Proc Natl Acad Sci U S A* **105**, 16731-6.

Strater, J., Wellisch, I., Riedl, S., Walczak, H., Koretz, K., Tandara, A., Krammer, P. H. and Moller, P. (1997). CD95 (APO-1/Fas)-mediated apoptosis in colon epithelial cells: a possible role in ulcerative colitis. *Gastroenterology* **113**, 160-7.

Strober, W., Fuss, I. J. and Blumberg, R. S. (2002). The immunology of mucosal models of inflammation. *Annu Rev Immunol* **20**, 495-549.

Su, L., Shen, L., Clayburgh, D. R., Nalle, S. C., Sullivan, E. A., Meddings, J. B., Abraham, C. and Turner, J. R. (2009). Targeted epithelial tight junction dysfunction causes immune activation and contributes to development of experimental colitis. *Gastroenterology* **136**, 551-63.

Sudo, N., Sawamura, S., Tanaka, K., Aiba, Y., Kubo, C. and Koga, Y. (1997). The requirement of intestinal bacterial flora for the development of an IgE production system fully susceptible to oral tolerance induction. *J Immunol* **159**, 1739-45.

Suzuki, K., Kawauchi Y, Palaniyandi SS, Veeraveedu PT, Fujii M, Yamagiwa S, Yoneyama H, Han GD, Kawachi H, Okada Y, Ajioka Y, Watanabe K, Hosono M, Asakura H, Aoyagi Y, Narumi S. (2007). Blockade of interferon-gamma-inducible protein-10 attenuates chronic experimental colitis by blocking cellular trafficking and protecting intestinal epithelial cells. *Pathol Int* **57**, 413-20.

Tannock, G. W. (2008). Molecular analysis of the intestinal microflora in IBD. *Mucosal Immunol* **1 Suppl 1**, S15-8.

Tap, J., Mondot, S., Levenez, F., Pelletier, E., Caron, C., Furet, J. P., Ugarte, E., Munoz-Tamayo, R., Paslier, D. L., Nalin, R. et al. (2009). Towards the human intestinal microbiota phylogenetic core. *Environ Microbiol* **11**, 2574-84.

Thompson, G. R. and Trexler, P. C. (1971). Gastrointestinal structure and function in germ-free or gnotobiotic animals. *Gut* **12**, 230-5.

Tilsala-Timisjärvi, A., Alatossava T,. (1997). Development of oligonucleotide primers from the 16S-23s rRNA intergenic sequences for identifying different dairy and probiotic lactic acid bacteria by PCR. *International Journal of Food Microbiology* **35**, 49-56.

Tubelius, P., Stan, V. and Zachrisson, A. (2005). Increasing work-place healthiness with the probiotic Lactobacillus reuteri: a randomised, double-blind placebo-controlled study. *Environ Health* **4**, 25.

Tursi, A., Brandimarte, G., Giorgetti, G. M., Forti, G., Modeo, M. E. and Gigliobianco, A. (2004). Low-dose balsalazide plus a high-potency probiotic preparation is more effective than balsalazide alone or mesalazine in the treatment of acute mild-to-moderate ulcerative colitis. *Med Sci Monit* **10**, PI126-31.

Tysk, C., Lindberg, E., Jarnerot, G. and Floderus-Myrhed, B. (1988). Ulcerative colitis and Crohn's disease in an unselected population of monozygotic and dizygotic twins. A study of heritability and the influence of smoking. *Gut* **29**, 990-6.

Ukena, S. N., Singh, A., Dringenberg, U., Engelhardt, R., Seidler, U., Hansen, W., Bleich, A., Bruder, D., Franzke, A., Rogler, G. et al. (2007). Probiotic Escherichia coli Nissle 1917 inhibits leaky gut by enhancing mucosal integrity. *PLoS ONE* **2**, e1308.

Ulisse, S., Gionchetti, P., D'Alo, S., Russo, F. P., Pesce, I., Ricci, G., Rizzello, F.,

Helwig, U., Cifone, M. G., Campieri, M. et al. (2001). Expression of cytokines, inducible nitric oxide synthase, and matrix metalloproteinases in pouchitis: effects of probiotic treatment. *Am J Gastroenterol* **96**, 2691-9.

van Baarlen, P., Troost, F. J., van Hemert, S., van der Meer, C., de Vos, W. M., de Groot, P. J., Hooiveld, G. J., Brummer, R. J. and Kleerebezem, M. (2009). Differential NF-kappaB pathways induction by Lactobacillus plantarum in the duodenum of healthy humans correlating with immune tolerance. *Proc Natl Acad Sci U S A* **106**, 2371-6.

Van den Steen, P. E., Husson, S. J., Proost, P., Van Damme, J. and Opdenakker, G. (2003). Carboxyterminal cleavage of the chemokines MIG and IP-10 by gelatinase B and neutrophil collagenase. *Biochem Biophys Res Commun* **310**, 889-96.

van der Kleij, H. P., O'Mahony, C., Shanahan, F., O'Mahony, L. and Bienenstock, J. (2008). Protective effects of Lactobacillus reuteri and Bifidobacterum infantis in murine models for colitis do not involve the vagus nerve. *Am J Physiol Regul Integr Comp Physiol*.

Van Gossum, A., Dewit, O., Louis, E., de Hertogh, G., Baert, F., Fontaine, F., DeVos, M., Enslen, M., Paintin, M. and Franchimont, D. (2007). Multicenter randomized-controlled clinical trial of probiotics (Lactobacillus johnsonii, LA1) on early endoscopic recurrence of Crohn's disease after Ileo-caecal resection. *Inflamm Bowel Dis* **13**, 135-42.

Venturi, A., Gionchetti, P., Rizzello, F., Johansson, R., Zucconi, E., Brigidi, P., Matteuzzi, D. and Campieri, M. (1999). Impact on the composition of the faecal flora by a new probiotic preparation: preliminary data on maintenance treatment of patients with ulcerative colitis. *Aliment Pharmacol Ther* **13**, 1103-8.

Vetrano, S., Correale, C., Borroni, E. M., Pagano, N., Savino, B., Locati, M., Malesci, A., Repici, A. and Danese, S. (2008). Colifagina, a novel preparation of 8 lysed bacteria ameliorates experimental colitis. *Int J Immunopathol Pharmacol* **21**, 401-7.

Vidal, K., Grosjean, I., evillard, J. P., Gespach, C. and Kaiserlian, D. (1993). Immortalization of mouse intestinal epithelial cells by the SV40-large T gene. Phenotypic and immune characterization of the MODE-K cell line. *J Immunol Methods* **166**, 63-73.

Wehkamp, J., Harder, J., Wehkamp, K., Wehkamp-von Meissner, B., Schlee, M., Enders, C., Sonnenborn, U., Nuding, S., Bengmark, S., Fellermann, K. et al. (2004). NF-kappaB- and AP-1-mediated induction of human beta defensin-2 in intestinal epithelial cells by Escherichia coli Nissle 1917: a novel effect of a probiotic bacterium. *Infect Immun* **72**, 5750-8.

Wehkamp, J., Salzman, N. H., Porter, E., Nuding, S., Weichenthal, M., Petras, R. E., Shen, B., Schaeffeler, E., Schwab, M., Linzmeier, R. et al. (2005a). Reduced Paneth cell alpha-defensins in ileal Crohn's disease. *Proc Natl Acad Sci U S A* **102**, 18129-34.

Wehkamp, J., Schmid, M., Fellermann, K. and Stange, E. F. (2005b). Defensin deficiency, intestinal microbes, and the clinical phenotypes of Crohn's disease. *J Leukoc Biol* **77**, 460-5.

Weiner, H. L. (1994). Oral tolerance. *Proc Natl Acad Sci U S A* **91**, 10762-5.

Wigg, A. J., Roberts-Thomson, I. C., Dymock, R. B., McCarthy, P. J., Grose, R. H. and Cummins, A. G. (2001). The role of small intestinal bacterial overgrowth, intestinal permeability, endotoxaemia, and tumour necrosis factor alpha in the pathogenesis of non-alcoholic steatohepatitis. *Gut* **48**, 206-11.

Winslet, M. C., Allan, A., Poxon, V., Youngs, D. and Keighley, M. R. (1994). Faecal diversion for Crohn's colitis: a model to study the role of the faecal stream in the inflammatory process. *Gut* **35**, 236-42.

Woodmansey, E. J. (2007). Intestinal bacteria and ageing. *J Appl Microbiol* **102**, 1178-86.

Wu, T. C. and Chen, P. H. (2009). Health consequences of nutrition in childhood and early infancy. *Pediatr Neonatol* **50**, 135-42.

Yan, F., Cao, H., Cover, T. L., Whitehead, R., Washington, M. K. and Polk, D. B. (2007). Soluble proteins produced by probiotic bacteria regulate intestinal epithelial cell survival and growth. *Gastroenterology* **132**, 562-75.

Yu, Y., Zeng, H., Lyons, S., Carlson, A., Merlin, D., Neish, A. S. and Gewirtz, A. T. (2003). TLR5-mediated activation of p38 MAPK regulates epithelial IL-8 expression via posttranscriptional mechanism. *Am J Physiol Gastrointest Liver Physiol* **285**, G282-90.

Zhao, L. and Ackerman, S. L. (2006). Endoplasmic reticulum stress in health and disease. *Curr Opin Cell Biol* **18**, 444-52.

Zyrek, A. A., Cichon, C., Helms, S., Enders, C., Sonnenborn, U. and Schmidt, M. A. (2007). Molecular mechanisms underlying the probiotic effects of Escherichia coli Nissle 1917 involve ZO-2 and PKCzeta redistribution resulting in tight junction and epithelial barrier repair. *Cell Microbiol* **9**, 804-16.

ACKNOWLEDGEMENTS

Doing a PhD is a challenge characterized by the alternation of ups and downs. As the latter ones are unevitable due to the complexity and variability of biological systems, the constant scientfic input as well as mental support of a lot of people was very important for the successful finalization of my PhD project.

First of all, special thanks go to my supervisor Prof. Dirk Haller, who contributed to this work by countless discussions and ideas, his never-ending enthusiasm as well as constant support and motivation during the last three years.

I also want to thank Prof. Siegfried Scherer for the time and energy to read this thesis as well as Prof. Michael Scheman for his agreement to act as a chairman for the present thesis.

I am deeply thankful to all my colleagues and friends at the Department for Biofunctionality for their scientific but especially for the mental support as well as for the fun we had, making work (and travel) a pleasure. My thanks go to the "old" ones, who went through a lot of ups and downs with me during the last three years: Micha Hoffmann, Eva Rath, Anja Messlik, Tanja Werner, Thomas Clavel and Katharina Rank. Shared gallows humor as well as a continuous supply of cakes and "Zopf" makes life much easier. My thanks also go to Christine Bäuerl and especially Marie Anne von Schillde, my companion in the "killing commando", who both contributed with a lot of work to the present thesis.
My thanks, also go to all the others who accompanied and supported me: Nico Gebhardt, Emanuel Berger, Sigrid Kisling, Benjamin Tielmann, Pia Baur, Natalie Steck, Job Mapesa, Stephan Wagner, Ingrid Schmöller, Susan Chang, Theresa Asen, Sylvia Pitariu, Katharina Heller, Lisa Gruber, Anna Zhenchuk, Pedro Ruiz, Anna Shkoda, Theresa Habermann, Sonja Böhm and Brita Sturm.

Apart from my colleagues at the Department of Biofunctionality there are several people from other departments and institutes who I would like to acknowledge for their important input on the present work: Gunnar Loh and Carl-Alfred Alpert from the German Insitute of Human Nutrition, Caroline Reiff and Margret Delday from the Rowett Institute for Nutrition and Health as well as Hannes Hahne from the Department of Bioanalytics.

Ich danke allen meinen Freunden für ihre Unterstützung und für ihr Verständnis für „was immer Du da eigentlich machst". Eure Nachfragen, wann ich denn endlich fertig bin, circa ab Jahr eins der Doktorarbeit, waren mir ein ständiger Ansporn. Die endlosen Telefon-und Bargespräche mit meinen übermotivierten „Leidensgenossinnen" Anya und Yasmine haben ihren ganz speziellen Beitrag zum Gelingen dieser Arbeit geleistet.

Mein herzlichster Dank gilt meinen Eltern Klaus und Marianne Hörmannsperger, meiner Schwester Nicole und Berni dafür, dass sie immer für mich da sind, mich anspornen und ganz einfach mein Leben bereichern.

Curriculum Vitae

Personal

name: Hörmannsperger Gabriele

date of birth: 25/06/1982

place of birth: Landshut, Germany

University

Since 06/2006   PhD student at the department for biofunctionality,
                         Technische Universität München

10/2004 until 05/2006   Master studies in Molecular Biotechnology
                         Technische Universität München

10/2001 until 09/2004   Bachelor studies in Molecular Biotechnology
                         Technische Universität München

Internship

09/2005 until 10/2005   University of Alberta, Edmonton, Canada
                         Experimental study in the group of Prof. Karen Madsen

Membership

Since 2006   Membership of the European Nutrigenomics Organisation,
                         Focus Team "Microbe-Host Interaction"